Augmented Vision and Reality

Volume 1

Series Editors

Riad I. Hammoud, DynaVox Technologies, Pittsburgh, PA, USA

Lawrence B. Wolff, Equinox Corporation, New York, USA

For further volumes:
http://www.springer.com/series/8612

Riad Hammoud · Guoliang Fan
Robert W. McMillan · Katsushi Ikeuchi
Editors

Machine Vision Beyond Visible Spectrum

Springer

Editors
Dr. Riad Hammoud
DynaVox Mayer-Johnson
Wharton Street 2100
Pittsburgh PA 15203
USA
e-mail:
Riad.Hammoud@dynavoxtech.com

Robert W. McMillan
U.S. Army Space and Missile Defense
Command
PO Box 1500
Huntsville AB 35807-3801
USA
e-mail: bob.mcmillan@us.army.mil

Guoliang Fan
School of Electrical and Computer
Engineering
Oklahoma State University
202 Engineering South
Stillwater OK
USA
e-mail: guoliang.fan@oksate.edu

Dr. Katsushi Ikeuchi
Institute of Industrial Science
University of Tokyo
Komaba 4-6-1
Meguro-ku, Tokyo
153-8505 Japan
e-mail: ki@cvl.iis.u-tokyo.ac.jp

ISSN 2190-5916
e-ISSN 2190-5924
ISBN 978-3-642-11567-7
e-ISBN 978-3-642-11568-4
DOI 10.1007/978-3-642-11568-4
Springer Heidelberg Dordrecht London New York

Cover design: eStudio Calamar, Berlin/Figueres

Printed on acid-free paper

Springer is part of Springer Science+Business Media (www.springer.com)

Preface

The genesis of this book on "Machine Vision Beyond the Visible Spectrum" is the successful series of seven workshops on Object Tracking and Classification Beyond the Visible Spectrum (OTCBVS) held as part of the IEEE annual Conference on Computer Vision and Pattern Recognition (CVPR) from 2004 through 2010. Machine Vision Beyond the Visible Spectrum requires processing data from many different types of sensors, including visible, infrared, far infrared, millimeter wave, microwave, radar, and synthetic aperture radar sensors. It involves the creation of new and innovative approaches to the fields of signal processing and artificial intelligence. It is a fertile area for growth in both analysis and experimentation and includes both civilian and military applications. The availability of ever improving computer resources and continuing improvement in sensor performance has given great impetus to this field of research. The dynamics of technology "push" and "pull" in this field of endeavor have resulted from increasing demand from potential users of this technology including both military and civilian entities as well as needs arising from the growing field of homeland security. Military applications in target detection, tracking, discrimination, and classification are obvious. In addition to this obvious use, Machine Vision Beyond the Visible Spectrum is the basis for meeting numerous security needs that arise in homeland security and industrial scenarios. A wide variety of problems in environmental science are potentially solved by Machine Vision, including drug detection, crop health monitoring, and assessment of the effects of climate change.

This book contains 10 chapters, broadly covering the subfields of *Tracking and Recognition in the Infrared, Multi-Sensor Fusion and Smart Sensors*, and *Hyperspectral Image Analysis*. Each chapter is written by recognized experts in the field of machine vision, and represents the very best of the latest advancements in this dynamic and relevant field.

The first chapter entitled "Local Feature Based Person Detection and Tracking Beyond the Visible Spectrum", by Kai Jüngling and Michael Arens of FGAN-FOM in Germany, addresses the very relevant topic of person detection and tracking in infrared image sequences. The viability of this approach is demonstrated by person detection and tracking in several real world scenarios.

"Appearance Learning by Adaptive Kalman Filters for Robust Infrared Tracking" by Xin Fan, Vijay Venkataraman and Joseph Havlicek of Oklahoma State University, Dalian Institute of Technology, and The University of Oklahoma, casts the tracking problem in a co-inference framework, where both adaptive Kalman filtering and particle filtering are integrated to learn target appearance and to estimate target kinematics in a sequential manner. Experiments show that this approach outperforms traditional approaches with near-super-pixel tracking accuracy and robust handling of occlusions. Chapter 3, "3D Model-Driven Vehicle Matching and Recognition", by Tingbo Hou, Sen Wang, and Hong Qin of Stony Brook University, treats the difficult and universal problem of vehicle recognition in different image poses under various conditions of illumination and occlusion. A compact set of 3D models is used to represent basic vehicle types, and pose transformations are estimated by using approximated vehicle models that can effectively match objects under large viewpoint changes and partial occlusions. Experimental results demonstrate the efficacy of this approach with the potential for extending these methods to other types of objects. The title of Chap. 4 is "Pattern Recognition and Tracking in Infrared Imagery" by Mohammad Alam of the University of South Alabama. This chapter discusses several target detection and tracking algorithms and compares the results obtained to real infrared imagery to verify the effectiveness of these algorithms for target detection and tracking. Chapter 5 describes "A Bayesian Method for Infrared Face Recognition" by Tarek Elguebaly and Nizar Bouguila of Concordia University. It addresses the difficult problem of face recognition under varying illumination conditions and proposes an efficient Bayesian unsupervised algorithm for infrared face recognition, based on the Generalized Gaussian Mixture Model.

Chapter 6, entitled "Fusion of a Camera and Laser Range Sensor for Vehicle Recognition", by Shirmila Mohottala, Shintaro Ono, Masataka Kagesawa, and Katsushi Ikeuchi of the University of Tokyo, combines the spatial localization capability of the laser sensor with the discrimination capability of the imaging system. Experiments with this combination give a detection rate of 100 percent and a vehicle type classification rate of 95 percent. Chapter 7 presents "A System Approach to Adaptive Multimodal Sensor Designs", by Tao Wang, Zhigang Zhu, Robert S. Krzaczek and Harvey E Rhody of the City College of New York, based on the integration of tools for the physics-based simulation of complex scenes and targets, sensor modeling, and multimodal data exploitation. The result of this work is an optimized design for the peripheral-fovea structure and a system model for developing sensor systems that can be developed within a simulation context.

Chapter 8, entitled "Statistical Affine Invariant Hyperspectral Texture Descriptors Based on Harmonic Analysis" by Pattaraporn Khuwuthyakorn, Antonio Robles-Kelly, and Jun Zhou of the Cooperative Research Centre for National Plant Biosecurity in Australia, focuses on the problem of recovering a hyperspectral image descriptor based on harmonic analysis. This chapter illustrates the robustness of these descriptors to affine transformations and shows their utility for purposes of recognition. "Tracking and ID via Object Reflectance Using a Hyperspectral Video Camera" is the title of Chap. 9. This chapter is authored by

Hien Nguyen, Amit Banerjee, Phil Burlina, and Rama Chellappa of the University of Maryland and focuses on the problem of tracking objects through challenging conditions, such as rapid illumination and pose changes, occlusions, and in the presence of confusers. This chapter demonstrates that the near-IR spectra of human skin can be used to distinguish different people in a video sequence. The final chapter of this book, "Moving Object Detection and Tracking in Forward Looking Aerial Imagery", by Subhabrata Bhattacharya, Imran Saleemi, Haroon Idrees, and Mubarak Shah of the University of Central Florida, discusses the challenges of automating surveillance and reconnaissance tasks for infrared visual data obtained from aerial platforms. This chapter gives an overview of these problems and the associated limitations of some of the conventional techniques typically employed for these applications.

Although the inspiration for this book was the OTCVBS workshop series, the subtopics and chapters contained herein are based on new concepts and new applications of proven results, and not necessarily limited to IEEE OTCBVS workshop series materials. The authors of the various chapters in this book were carefully chosen from among practicing application-oriented research scientists and engineers. All authors work with the problems of machine vision or related technology on a daily basis, and all are internationally recognized as technical experts in the fields addressed by their chapters.

It is the profound wish of the editors and authors of this book that it will be of some use to practicing scientists and engineers in the field of machine vision as they endeavor to improve the systems on which so many of us rely for safety and security.

June 2010

Riad Hammoud
Guoliang Fan
Robert W. McMillan
Katsushi Ikeuchi

Contents

Part I
Tracking and Recognition in Infrared

Local Feature Based Person Detection and Tracking Beyond the Visible Spectrum

Kai Jüngling and Michael Arens

Abstract One challenging field in computer vision is the automatic detection and tracking of objects in image sequences. Promising performance of local features and local feature based object detection approaches in the visible spectrum encourage the application of the same principles to data beyond the visible spectrum. Since these dedicated object detectors neither make assumptions on a static background nor a stationary camera, it is reasonable to use these object detectors as a basis for tracking tasks as well. In this work, we address the two tasks of object detection and tracking and introduce an integrated approach to both challenges that combines bottom-up tracking-by-detection techniques with top-down model based strategies on the level of local features. By this combination of detection and tracking in a single framework, we achieve (i) automatic identity preservation in tracking, (ii) a stabilization of object detection, (iii) a reduction of false alarms by automatic verification of tracking results in every step and (iv) tracking through short term occlusions without additional treatment of these situations. Since our tracking approach is solely based on local features it works independently of underlying video-data specifics like color information—making it applicable to both, visible and infrared data. Since the object detector is trainable and the tracking methodology does not make any assumptions on object class specifics, the overall approach is general applicable for any object class. We apply our approach to the task of person detection and tracking in infrared image sequences. For this case we show that our local feature based approach inherently

K. Jüngling (✉) · M. Arens
Fraunhofer IOSB, Gutleuthausstrasse 1 76275 Ettlingen, Germany
e-mail: kai.juengling@iosb.fraunhofer.de

M. Arens
e-mail: michael.arens@iosb.fraunhofer.de

R. Hammoud et al. (eds.), *Machine Vision Beyond Visible Spectrum*,
Augmented Vision and Reality, 1, DOI: 10.1007/978-3-642-11568-4_1,

allows for object component classification, i.e., body part detection. To show the usability of our approach, we evaluate the performance of both, person detection and tracking in different real world scenarios, including urban scenarios where the camera is mounted on a moving vehicle.

Keywords Person detection · Person tracking · Visual surveillance · SURF

1 Introduction

Object, and specifically person or pedestrian detection and tracking has been subject to extensive research over the past decades. The application areas for this are vast and reach from video surveillance, thread assessment in military applications, driver assistance to human computer interaction. An extensive review of the whole field of pedestrian detection and tracking is beyond the scope of this paper and can be found in [11, 18, 40]. We will indicate, however, some representative work for each of what we think to be escalating levels of difficulty: (i) person detection, (ii) person tracking and (iii) person detection and tracking from moving cameras.

Early systems in person centered computer vision applications mainly focused on surveillance tasks with stationary cameras. Here, full systems like [16, 37] built on foreground detection methods that model the static background and detect persons as foreground regions. These methods [33] have been extensively studied and improved over the years. Some research in this area has focused on this topic for the specific case of thermal imagery [7, 10], while some research fuses information from infrared and the visible spectrum [9, 27]. Drawbacks of systems that rely on person detection by foreground segmentation are the disability to reliably distinguish different object classes and to cope with ego-motion of the recording camera, though extensions in this latter direction have been proposed by [5, 31]. Both problems can be solved by using a dedicated object detector to find people in images.

Recent advances in object detection in the visible spectrum [8, 13, 24, 32, 36, 38] encourage the application of these trainable, class-specific object detectors to thermal data. Although person detection in infrared has its own advantages as well as disadvantages when compared to detection in the visible spectrum [12], most principles can be transferred from the visible spectrum to infrared. While some techniques like the Histogram of Oriented Gradients (HOG) [8] can be directly be transferred to infrared data [34], a lot of research focuses specifically on person detection in infrared. Nanda and Davis [30] use a template based approach which builds on training samples of persons to detect person in infrared data. In [39], Xu and Fujimura use a SVM which also builds on size normalized person samples to detect and track persons. In [6], Bertozzi et al. detect pedestrians from a moving vehicle by localization of symmetrical objects with specific size and aspect ratio, combined with a set of matches filters.

For most high level applications like situation assessment, the person detection results alone are not sufficient since they only provide a snapshot of a single point in time. For these higher level interpretation purposes, meaningful person trajectories have to be built by a tracking process. To benefit from the advantages of the dedicated object detectors, a lot of approaches directly built on the results of these person detectors to conduct tracking: Andriluka et al. introduced a method of combining tracking and detection of people in [1]. This approach uses knowledge of the walking cycle of a person to predict a persons position and control the detection. Another extension of this work [24] was proposed in [26] where a tracking was set up on the ISM based object detector. In [25] Leibe et al. further extended the work to track people from a moving camera. Gammeter et al. [14] built the tracking based on the object detector and additional depth cues obtained from a stereo camera to track people in street scenes from a moving camera. In [15], Gavrila and Munder proposed a multi cue pedestrian detection and tracking system that is applicable from a moving vehicle too. They use a cascade of detection modules that involves complementary information including stereo. Wu and Nevatia [38] introduced a system that detects body parts by a combination of edgelet features and combines the responses of the part detectors to compute the likelihood of the presence of a person. The tracking is conducted by a combination of associating detection results to trajectories and search for persons with mean shift. In both cases, an appearance model which is based on color is used for data association in tracking.

In infrared data, person tracking is a more challenging problem than in the visible spectrum. This is due to similar appearance of persons in infrared which makes identity maintenance in tracking much more difficult compared to the visible spectrum where rich texture and color is available to distinguish persons. Especially on moving cameras, where the image position of people is unstable and thus not sufficient to correctly maintain object identities, the above mentioned approaches would not be capable to track persons robustly. This is due to the different assumptions the approaches make on the availability of color, a stationary camera or special sensors like a stereo camera. An approach which focuses on pedestrian tracking without making these assumption is presented in [39] by Xu and Fujimura. Here, the tracking is built on the infrared person detection results of the SVM classifier. For that they use a Kalman filter to predict a persons position and combine this with a mean shift tracking.

In this chapter, we seize on the task of detecting and tracking multiple objects in real-world environments from a possibly moving, monocular infrared camera and by that pursue the work presented in [20, 21]. Although we focus on detecting and tracking people, our approach works independently of object specifics and is thus generically applicable for tracking any object class.

Unlike most of the before mentioned approaches we do not make any assumptions on application scenario, environment or sensor specifics. Our whole detection and tracking approach is solely built on local image features (see [35] for an extensive overview) which are perfectly suited for this task since they are available in every sensor domain. As local features, we picked SURF [2] (replaceable with SIFT [28]) features since, in our application, they have some major advantages compared to other local features like a combination of Harris keypoints [17] and shape descriptors [3] (as used in [23]).

On the keypoint level, SURF features respond to blob-like structures rather than to edges, which makes them well suited for infrared person detection since people here appear as lighter blobs on darker background (or inverted, dependent on sensor data interpretation). This is due to the use of a hessian matrix based keypoint detector (Difference of Gaussian which approximates the Laplacian of Gaussian in case of SIFT) which responds to blobs rather than to corners and edges like, e.g., Harris based keypoint detectors. The SURF descriptor is able to capture two things which are important in detection and tracking. It captures the shape of a region which is important in the training of the general person detector, because the shape of person is alike for all people. Second, it is able to capture texture (which still might be available despite infrared characteristics) properties of the regions which is important in tracking where different persons have to be distinguished from each other. Another important property is the ability of the descriptor to distinguish between light blobs on dark background and dark blobs on light background. This makes it perfectly suited for detecting people in thermal data because those here usually appear lighter than the background (or darker, dependent on sensor data interpretation).

Our detection approach is built on the Implicit Shape Model (ISM) based approach introduced in [24]. Here, a general appearance codebook is learned based on training samples. Additionally to just detecting persons as a compound, we show how this local feature based person detector can be used to classify a person's body parts, which can be input to further articulation interpretation approaches. For tracking, we introduce a novel technique that is directly integrated into the ISM based detection and needs no further assumptions on the objects to be tracked. Here, we unite object tracking and detection in a single process and thereby address the tracking problem while enhancing the detection performance. The coupling of tracking and detection is carried out by a projection of expectations resulting from tracking into the detection on the feature level. This approach is suited to automatically combine new evidence resulting from sensor data with expectations gathered in the past. By that, we address the major problems that exist in tracking: we automatically preserve object identity by integrating expectation into detection, and, by using the normal codebook-matching procedure, we automatically integrate new data evidence into existing hypotheses. The projection of expectation thus stabilizes detection itself and reduces the problem of multiple detections generated by a single real world object. By adapting the weights of projected features over time, we automatically take the history and former reliability of a hypothesis into account and therefore get by without a special approach to assess the reliability of a tracked hypothesis. Using this reliability assessment, tracks are automatically initialized and terminated in detection.

We evaluate both, the standalone person detector and the person tracking approach. The person detector is evaluated in three thermal image sequences with a total of 2,535 person occurrences. These image sequences cover the complete range of difficulties in person detection, i.e., people appearing at different scales, visible from different viewpoints, and occluding each other. The person tracking is evaluated in these three and two additional image sequences under two main aspects. First, we show how tracking increases detection performance in the first three image sequences. Second we show how our approach is able to perform tracking in difficult situations where people move beside each other and the camera is moving. Additionally, we show that the tracking is even able to track people correctly in cases where strong camera motion occurs.

This chapter is structured as follows. Section 2 covers the standalone person detection. Here, we start by introducing the detection approach in Sect. 2.1. The body part classification is described in Sect. 2.2. Section 2.3 provides an evaluation of person detection. Person tracking is discussed in Sect. 3. This section includes the introduction of our tracking approach in Sect. 3.1, the tracking evaluation in Sect. 3.2 and the tackling of strong camera motion in tracking in Sect. 3.3. Section 4 closes this chapter with a conclusion.

2 Person Detection

This section focuses on person detection. It introduces the detection technique, shows how this can be employed to classify a person's body parts and presents experimental results.

2.1 Local Feature Based Person Detection

The person detection approach we use here is based on the trainable ISM object detection approach introduced in [24]. In this section, we briefly describe the training and detection approach and the enhancements we made.

2.1.1 Training

In the training stage, a specific object class is trained on the basis of annotated sample images of the desired object category. The training is based on local features that are employed to build an appearance codebook of a specific object category.

The SURF features extracted from the training images on multiple scales are used to build an object category model. For that purpose, features are first clustered in descriptor space to identify reoccurring features that are characteristic for the specific object class. To generalize from the single feature appearance and build a generic, representative object class model, the clusters are represented by the cluster center (in descriptor space). At this point, clusters with too few contributing features are removed from the model since these cannot be expected to be representative for the object category. The feature clusters are the basis for the generation of the Implicit Shape Model (ISM) that describes the spatial configuration of features relative to the object center (see Fig. 1a) and is used to vote for object center locations in the detection process. This ISM is built by comparing every training feature to each

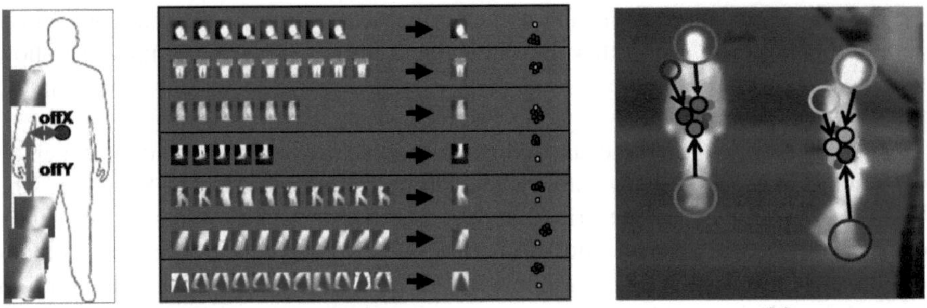

(a) (b) (c)

Fig. 1 **a** ISM describes spatial configuration of features relative to object center. **b** Clustered training features are mapped to a prototype. Each codebook entry contains a prototype and the spatial distribution of features. **c** Image features that match to codebook prototypes cast votes for object center locations. Each image feature has only a single vote in the final detection set since a single image feature can only provide evidence for one object hypothesis

prototype (cluster center) that was generated in the previous clustering step. If the similarity (euclidean distance in descriptor space) of a feature and the prototype is above an assignment threshold, the feature is added to the specific codebook entry. Here, the feature position relative to the object center—the offset—is added to the spatial distribution of the codebook (Fig. 1b) entry with an assignment probability. This probability is based on descriptor similarity and a single feature can contribute to more than one codebook entry (fuzzy assignment).

2.1.2 Detection

To detect objects of the trained class in an input image, again SURF features are extracted. These features (the descriptors) are then matched with the codebook, where codebook entries with a distance below a threshold t_{sim} are activated and cast votes for object center locations (Fig. 1c). To allow for fast identification of promising object hypothesis locations, the voting space is divided into a discrete grid in x-, y-, and scale-dimension. Each grid that defines a voting maximum in a local neighborhood is taken to the next step, where voting maxima are refined by mean shift to accurately identify object center locations.

At this point we make two extensions to the work of [24]. First, we do not distribute vote weights equally over all features and codebook entries but use feature similarities to determine the assignment probabilities. By that, features which are more similar to codebook entries have more influence in object center voting. The assignment strength $p(C_i \mid f_k)$ of an image feature f_k, codebook entry C_i combination is determined by:

$$p(C_i \mid f_k) = \frac{t_{\mathrm{sim}} - \rho(f_k, C_i)}{t_{\mathrm{sim}}}, \tag{1}$$

where $\rho(f_k, C_i)$ is the euclidean distance in descriptor space. Since all features with a distance above or equal t_{sim} have been rejected before, $p(C_i|f_k)$ is in range [0, 1]. The maximal assignment strength 1 is reached when the euclidean distance is 0. The same distance measure is used for the weight $p(V_{\vec{x}}|C_i)$ of a vote for an object center location $\vec{x}$ when considering a codebook entry C_i. The vote location $\vec{x}$ is determined by the ISM that was learned in training. Here, $\rho(f_k, C_i)$ is the similarity between a codebook prototype and a training feature that contributes to the codebook entry. The overall weight of a vote $V_{\vec{x}}$ is:

$$V_{\vec{x}}^{w} = p(C_i|f_k)p(V_{\vec{x}}|C_i). \tag{2}$$

Second, we approach the problem of the training data dependency. The initial approach by Leibe et al. uses all votes that contributed to a maximum to score a hypothesis and to decide which hypotheses are treated as objects and which are discarded. As a result, the voting and thus the hypothesis strength depends on the amount and character of training data. Features that frequently occurred in training data generate codebook entries that comprise many offsets. A single feature (in detection) that matches with the codebook prototype thus casts many votes in object center voting with the evidence of only a single image feature. Since a feature count independent normalization is not possible at this point, this can result in false positive hypotheses with a high score, generated by just a single or very few false matching image features. To solve this issue, we only count a single vote—the one with the highest similarity of image and codebook feature—for an image feature/hypothesis combination (see Fig. 1c). We hold this approach to be more plausible since a single image feature can only provide evidence for an object hypothesis once.

The score γ of a hypothesis ϕ can thus, without the need for a normalization, directly be inferred by the sum of weights of all I contributing votes:

$$\gamma_{\phi} = \sum_{i=1}^{I} V_i^{w}. \tag{3}$$

Certainly, this score is furthermore divided by the volume of the scale-adaptive search kernel (see [24] for details), which is necessary because objects at higher scales can be expected to generate much more features than those at lower scales. Additionally, this enhancement provides us with an unambiguousness regarding the training feature that created the involvement of a specific image feature in a certain hypothesis. This allows for decisive inference from a feature that contributed to an object hypothesis back to the training data. This is important for the classification of body parts which is described in detail in Sect. 2.2.

The result of the detection step is a set Φ of object hypotheses, each annotated with a score γ_{ϕ}. This score is subject to a further threshold application. All object hypotheses below that threshold are removed from the detection set Φ.

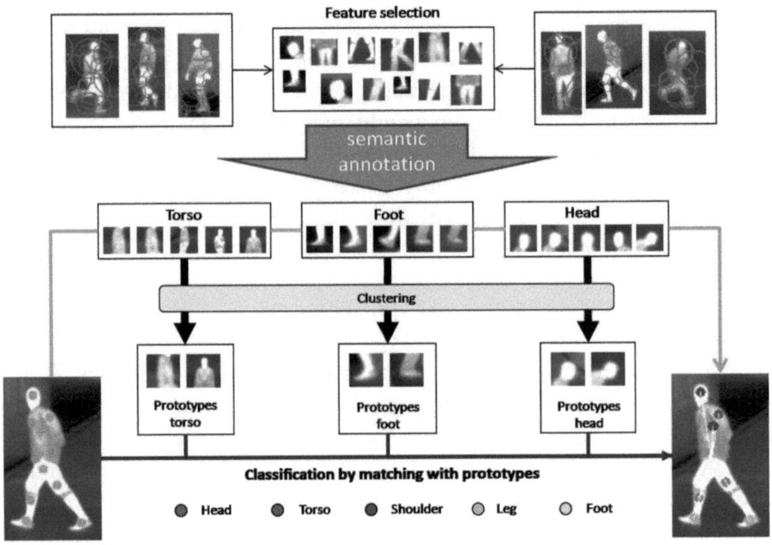

Fig. 2 Procedure of body part classification. Features found on body parts are annotated with the appropriate semantics, feature descriptors are then clustered to build appearance prototypes of each body part. Body part classification happens in two ways, the *top line* denotes the way of direct classification using the training annotation. The *bottom line* denotes classification by matching with the appearance prototypes

2.2 Body Part Classification

As mentioned in Sect. 2.1.2, our enhancements provide us with an unambiguousness regarding the training feature that created a specific vote. This unambiguous inference together with an object part annotation of the training data, i.e., a body part annotation of persons, allows for object-part classification. The training data body part annotation can directly be used to annotate training features found on body parts with semantic body part identifiers. This annotation is added to codebook entries for features that can be associated with certain body parts. Object hypotheses resulting from detection consist of a number of votes. The votes were generated by specific offsets (which refer to training features) in certain codebook entries which were activated by image features. As outlined in Fig. 2, using the annotation of these entries, we are now able to infer the semantics of image features that contribute to an object hypothesis.

This body part classification approach has the weakness that the similarity between an image feature and the training feature is calculated only indirectly by the similarity between the codebook representative and the image feature (see Eq. 1). This means that a feature that is annotated with a body part and resides in a specific codebook entry could contribute to a person hypothesis because the similarity between an image feature and the codebook representative is high enough (this similarity constraint is rather weak since we want to activate all similar structures for detection) but the image feature does in fact not represent the annotated body part.

For this reason, we decided to launch another classification level that includes stronger constraints on feature similarity and introduces a body part specific appearance generalization. Following that, we generate body part templates for every body part class found in training data, i.e., we pick all features annotated with "foot" from training data. The descriptors of these features are then clustered in descriptor space to generate body part templates. The presets on descriptor similarity applied here are stricter than those used in codebook training. This is because we rather want to generate an exact representation than to generalize too much from different appearances of certain body parts. The clustering results in a number of disjoint clusters that represent body parts. The number of descriptors in a cluster is a measure for how generic it represents a body part. The more often a certain appearance of a body part has been seen in training data, the more general this appearance is (since it was seen on many different people). Since the goal is to create an exact (strong similarity in clustering) and generic (repeatability of features) representation, we remove clusters with too few associated features. The remaining clusters are represented by their cluster center and constitute the templates. These templates can now be used to verify the body part classification of stage one by directly comparing the feature descriptors of a classified image feature with all templates of the same body part class. If a strong similarity constraint is met for any of the templates, the classification is considered correct. Otherwise, the image feature annotation is removed.

Example results of the body part classification are shown in Fig. 3. Here, the relevant body part categories are: head, torso, shoulder, leg, and foot. We see that we are not able to detect every relevant body part in any case, but the hints can be used—especially when considering temporal development—to build a detailed model of a person which can be the starting point for further interpretation of the person's articulation. (Compare [22] for work in this direction.)

2.3 Experimental Results

2.3.1 Training Data

A crucial point in the performance of a trainable object detector is the choice of training data. Our person detector is trained with a set of 30 training images taken from an image sequence that was acquired from a moving camera in urban terrain with a resolution of 640×480. The set contains eight different persons appearing at multiple scales and viewpoints. The persons are annotated with a reference segmentation which is used to choose relevant features to train the person detector. Additionally, we annotate the training features with body part identifiers when this is adequate

Fig. 3 Example body part classification results of detected persons. Relevant body part classes are: head, torso, shoulder, leg, and foot

(when a feature visually refers to a certain body part). Example results for the body part detection are shown in Fig. 3. All detection results shown hereafter do not contain any of the persons that appear in training data.

2.3.2 Person Detection

To show the operationality of the detection approach in infrared images, we evaluate the performance in three different image sequences, taken from different cameras under varying environmental conditions. For evaluation, all persons whose head or half of the body is visible are annotated with bounding boxes.

To assess the detection performance, we use the performance measure

$$recall = \frac{|true\ positives|}{|ground\ truth\ objects|} \tag{4}$$

following [25]. To determine whether an object hypothesis is a true- or a false positive, we use two different criteria. The *inside bounding box* criterion assesses an object hypothesis as true-positive if its center is located inside the ground truth bounding box. Only a single hypothesis is counted per ground truth object, all other hypotheses in the same box are counted as false positive. The *overlapping* criterion assesses object hypotheses using the ground truth and hypotheses bounding boxes. The overlap between those is calculated by the Jaccard-Index [19] (compare intersection-over-union criterion):

$$overlap = \frac{area(B_p \cap B_{gt})}{area(B_p \cup B_{gt})}. \tag{5}$$

The first criterion is deliberately used to account for inaccuracies in bounding boxes in the ground truth data and to assess the detection performance independently of its accuracy. Specifically in our case, where the bounding box is defined by the minimal box that contains all features which voted for a hypothesis, a hypothesis that only contains the upper body of a person would be counted as false positive using the overlapping criterion, even if all body parts of the upper body are correctly found. To depict the accuracy of detections, we use the overlapping criterion which is evaluated for different overlap demands.

The first image sequence contains a total of 301 person occurrences, appearing at roughly the same scale. People run from right to left in the camera's field of view with partial person–person overlapping. We evaluate the sequence using the recall criterion and the false positives per image. The recall is shown as a function of false positives per image as used in various object detector evaluations. To assess the accuracy of the detection we evaluate with different requirements of overlapping. The results for the different evaluation criteria (OLx: Bounding box overlap with a minimum overlap of $x\%$; BBI: Inside bounding box) are shown in Fig. 5a. The curves are generated by running the object detector with different parameter settings on the same image sequence. Example detections for this image sequence are shown in the top row of Fig. 4.

The second image sequence is from OTCBVS dataset [9] with 763 person occurrences. Here, a scene is observed by a static camera with a high-angle shot. Two persons appearing at a low scale move in the scene without any occlusions. As we see in Fig. 5b, the detection performance is very similar for all false positive rates. Here, we nearly detect all person occurrences in the image at low false positive rates. The results do not improve significantly with other parameters that allow person detections with lower similarity demands and result in more false positives. It is worth mentioning that the detector was trained on persons the appearance of which was not even close to the ones visible in this image sequence. Both, viewpoint and scale of the persons have changed completely between training and input data. Note that the buckling in the curves of bounding box overlap can result from parameter adjustment in allowed feature similarity for detection. Activating more image features for detection can result in more false positive hypotheses and in additional inaccuracies in the bounding box and thus in less true-positives regarding the overlap criterion. The detailed trend of false positives per image and recall for different overlap demands in Fig. 5d shows that the detection performance itself is very good. The accuracy is rather poor compared to the detection performance but still has a recall of above 0.7 with a 50% bounding-box overlap demand. With increasing overlap

Fig. 4 Example detections of all three test sequences. Sequence 1: *top row*, sequence 2: *middle row*, sequence 3: *bottom row*. *Dots* indicate features that generate the hypothesis marked with the *bounding box*

demand, the detection rate decreases and the false positives increase. As we can see from the development of the curves, this is just due to inaccuracy and not due to "real" false positives generated from background or other objects. Example detections for this image sequence are shown in the second row of Fig. 4.

The third image sequence was taken in urban terrain from a camera installed on a moving vehicle. This image sequence, with a total of 1,471 person occurrences, is the most challenging because a single image contains persons at various scales and the moving paths of persons cross, which leads to strong occlusions. From the example result images in the bottom row of Fig. 4, we see that some persons in the background occupy only few image pixels while other persons in the foreground take a significant portion of the whole image. Unlike one could expect, the fact that people are moving parallel to the camera is not very advantageous for the object detector because the persons limbs are not visible very well from this viewpoint. The results of this image sequence are shown in Fig. 5c. We see, that the *inside bounding box* criterion performs well and has a recall of more than 0.9 with less than 1.5 false positive/image. When applying the bounding box overlap criterion, the performance drops significantly—more than in image sequence one and two. Especially the 50% overlap criterion only reaches a recall of 0.5 with more than 5 false positives/image. This rapid performance degradation is mainly due to inaccuracies in bounding boxes

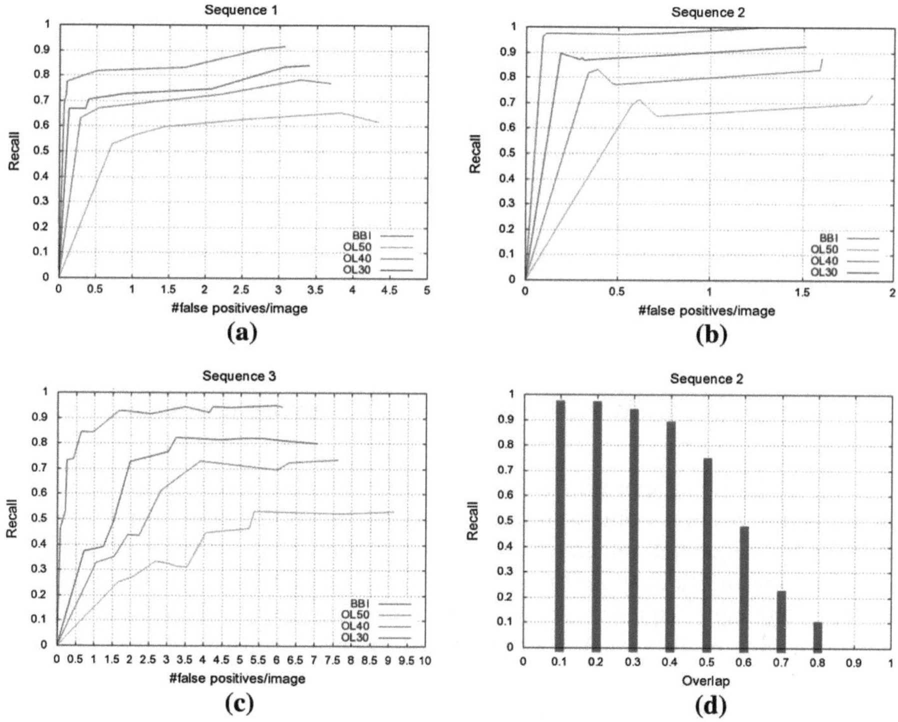

Fig. 5 Recall/false positive curves for **a** sequence 1, **b** sequence 2, and **c** sequence 3. Each chart contains four curves that refer to the different evaluation criteria. BBI: inside bounding box criterion. OL30/40/50: bounding box overlap criterion with 30, 40 and 50% overlap demand. **d** Trend of detection performance of sequence 2 with a single parameter set using different bounding box overlap demands (displayed on the x-axis in 10% steps)

of persons appearing at higher scales. This is also visible in the example detections in the bottom row of Fig. 4. Here, people in the scene background are most often detected accurately while persons close to the camera are detected rather imprecisely in terms of exact bounding boxes.

3 Person Tracking

Even a perfectly working person detector gives only a snapshot image of the surrounding. For most applications, like driver assistance or visual surveillance, it is necessary to interpret the situation over a time interval, i.e., to know where people are walking and thus know if they are a possible thread (spec. in military applications) or if we (as a driver of a vehicle) might be a thread to the person. For this, a person tracking is necessary. An important point in tracking is to consistently maintain object identities because this is a prerequisite for correct trajectory estimation. This is a difficult problem specifically in infrared data, where features like color that are commonly used to distinguish persons in tracking are not available. Here, people usually appear as a light region on a darker background which means the appearance of different persons is very alike. Additional difficulties arise when tracking should be conducted from a moving camera. In this case the use of position information for correct trajectory estimation is problematic since the camera motion distorts estimation of people motion.

In this section, we introduce a tracking strategy which is based on the object detector introduced in Sect. 2 and copes with the difficulties for tracking in infrared from a moving camera.

3.1 Local Feature Based Integration of Tracking and Detection

The object detection approach described up to now works exclusively data-driven by extracting features bottom-up from input images. At this point, we introduce a tracking technique that integrates expectations into this data-driven approach. The starting point of tracking are the results of the object detector applied to the first image of an image sequence. These initial object hypotheses build the basis for the object tracking in the sequel. Each of these hypotheses consists of a set of image features which generated the according detection. These features are employed to realize a feature based object-tracking.

3.1.1 Projection of Object Hypotheses

For every new image of the image sequence, all hypotheses Γ known in the system at this time T, each comprising a set of features Π_γ, are fed back to the object detection before executing the detection procedure. For the input image, the feature extraction is performed, resulting in a set of image features Π_{img}. For every object hypothesis in the system, the feature set Π_γ of this hypothesis γ is projected into the image. For that, we predict the feature's image positions for the current point in time (a Kalman-Filter that models the object-center dynamics assuming constant object acceleration is used to determine position prediction for features. Note that this is thought to be a weak assumption on object dynamics) and subjoin these feature to the image features.

In this joining, three different feature types are generated: The first feature type, the *native image features* Π_{img} refers to features that are directly extracted from the input image. These features contribute with the weight $P_{\text{type=nat}}$, which is set to 1.

The second feature type, the *native hypothesis features*, is generated by projecting the hypothesis features Π_γ to the image. These features are weighted with $P_{\text{type=hyp}}$ and are added to the detection-feature-set Π_γ^{tot} of hypothesis γ:

$$\Pi_\gamma^{\text{tot}} = \Pi_{\text{img}} \cup \Pi_\gamma . \tag{6}$$

These features integrate expectation into detection and their weight is set to a value in the range [0–1].

The next step generates the features of the third type, the *hypothesis features with image feature correspondence*. For this purpose, the hypothesis features Π_γ are matched (similarity is determined by an euclidean distance measure) with the image features Π_{img}. Since (i) the assignment of hypothesis to image features includes dependencies between assignments and since (ii) a single hypothesis feature can only be assigned to one image feature (and vice versa), a simple "best match" assignment is not applicable. We thus solve the feature assignment problem by the revised Hungarian method presented by Munkres in [29]. By that the best overall matching assignment and mutual exclusivity is ensured.

Feature assignments with a distance (in descriptor space) exceeding an assignment threshold κ_{feat} are prohibited. An additional image-distance constraint for feature pairs ensures the spatial consistency of features. Every $\iota \in \Pi_{\text{img}}$ which has a $\pi \in \Pi_\gamma$ assigned, is labeled as feature type 3 and contributes with the weight $P_{\text{type=mat}}$ (the matching hypothesis feature π is removed from the detection set: $\Pi_\gamma^{\text{tot}} = \Pi_\gamma^{\text{tot}} \setminus \pi$ to not count features twice). This weight is set to a value >1, because this feature type indicates conformity of expectation and data and thus contributes with the highest strength in the voting procedure.

The feature-type-weight is integrated into the voting by extending the vote weight (see Eq. 2) with factor P_{type} to

$$V_{\vec{x}}^{w} = p(C_i | f_k) \cdot p(V_{\vec{x}} | C_i) \cdot P_{\text{type}}. \tag{7}$$

The voting procedure—which is the essential point in object detection—is thus extended by integrating the three different feature types that contribute with different strengths. The whole procedure is shown in Fig. 6.

3.1.2 Coupled Tracking and Detection

From now on, the detection is executed following the general scheme described in Sect. 2. In addition to the newly integrated weight factor, the main difference to the standard detection is that the voting space contains some votes which vote exclusively for a specific object hypothesis. Besides, votes which were generated from native image features can vote for any hypothesis. This is shown in Fig. 7a. Here, different gray values visualize affiliation to different hypotheses.

Since the number and position of expected object hypotheses is known, no additional maxima search is necessary to search for known objects in the voting space. As we see in Fig. 7b, the mean shift search can be started immediately since the expected position of a hypothesis in voting space is known (the position is determined by a prediction using a Kalman filter that models object center dynamics). Starting from

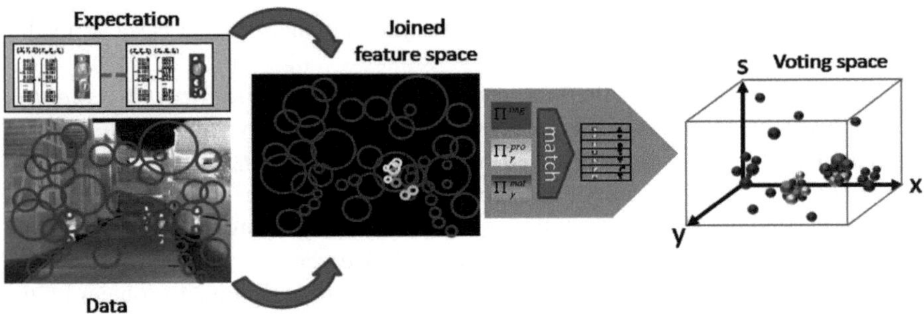

Fig. 6 Coupling of expectation and data for tracking. Features in object hypotheses (results of former detections) are propagated to the next frame and combined with new image features in a joined feature space. This feature space contains three different feature types which are generated by matching expectation and data. Π^{img}: native image features without correspondence in the hypothesis feature set, Π^{pro}: features of projected hypotheses without image feature match, Π^{mat}: matches of hypothesis and image features. The projected and matching features are marked with *grey values* according to the different hypotheses. These features can only vote for the hypotheses they refer to. The joined feature set is then input to the standard object detection approach where features are matched with the codebook to generate the voting space. Here, votes produced by native image features can vote for any object hypothesis while hypothesis specific votes are bound to a specific hypothesis

this position, the mean shift search is conducted determining the new object position. Since a mean shift search was started for every known object in particular, the search procedure knows which object it is looking for and thus only includes votes for this specific object and native votes into its search. By that hypothesis specific search, identity preservation is automatically included in the detection procedure without any additional strategy to assign detections to hypotheses. After mean shift execution, object hypotheses are updated with the newly gathered information. Since, by the propagation of the features, old and new information is already combined in the voting space, the object information in the tracking system can be replaced with the new information without any further calculations or matching.

To detect new objects, a search comprising the standard maximum search has to be conducted since the positions of new objects are not known beforehand. As we see in Fig. 7c, this maxima search is executed in a reduced voting space where only native votes that have not been assigned to a hypothesis yet remain. All votes that already contributed to an object hypothesis before are removed from the voting space. This ensures that no "double" object hypotheses are generated and determines that new image features are more likely assigned to existing object hypotheses than to new ones.

As in the original voting procedure, the initial "grid maxima" are refined with mean shift as we see in Fig. 7d. All maxima with a sufficient score found here initialize new tracks.

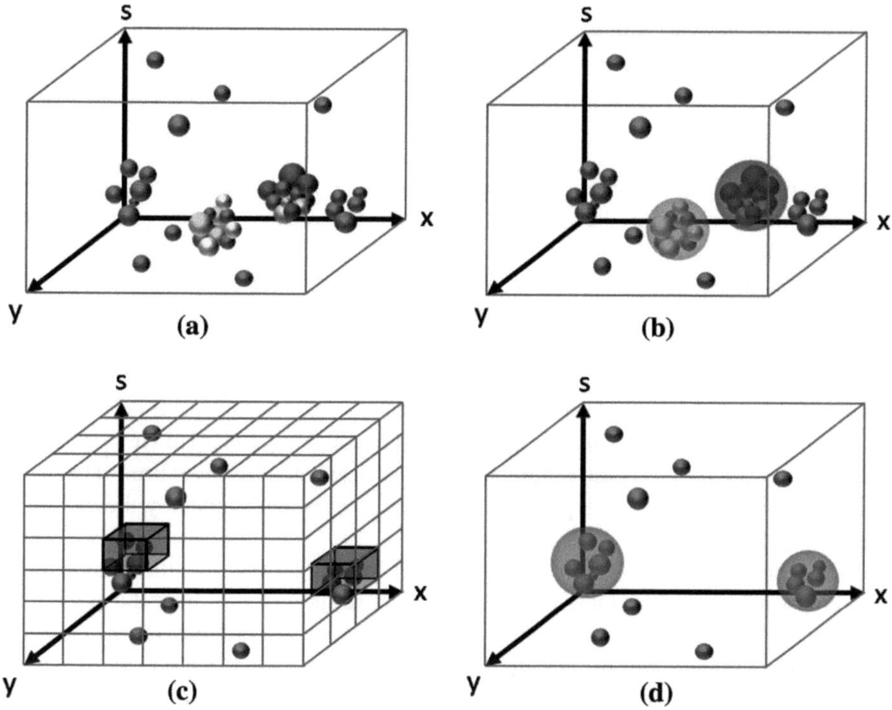

Fig. 7 Hypothesis search in the voting space. **a** Joined voting space with three vote types: native votes generated by image features without correspondence, votes generated by projected features without image feature correspondence, votes generated from hypothesis features with image feature correspondence. *Different grey values* visualize affiliation to object hypotheses. **b** Mean shift search for known object hypotheses. No initial maxima search is necessary since approximate positions of objects are known. **c** Grid maxima search to detect new objects in the reduced voting space. **d** Mean shift search to refine maxima positions of new objects

3.1.3 Inherent Reliability Adaption

A detection resulting from a projected hypothesis already contains the correctly updated information since the integration of old and new information has been conducted in the detection process itself. The reliability of this detection thus already reflects the hypothesis reliability with inclusion of the hypothesis history. This is due to the inclusion of projected features into the detection. To achieve automatic adaption of the reliability over time, we replace the constant factor P_{type} in Eq. 7 by a feature specific, time-varying function $P_{\text{type}}^{\pi,t}$ which is adapted every time the feature contributes to a detection. By that, feature history is inherently included. To accomplish this, the type factor $P_{\text{type}}^{\pi,t}$, of a feature $\pi \in \Pi_\gamma$ is set to

$$P_{\text{type}}^{\pi,t} = P_{\text{type}}^{\pi,t-1} \cdot \alpha_{\text{type}} \tag{8}$$

at time t. $P_{\text{type}}^{\pi,t-1}$ is the previous type-weight and α_{type} is the type-determined adaption rate previously used for the constant case (we use $\alpha_{\text{nat}} = 1$, $\alpha_{\text{hyp}} = 0.9$, and $\alpha_{\text{mat}} = 1.1$).

This rule leads to an automatic adaption of the feature weight determined by the presence of data evidence. Initially, all features have the same type weight 1 since they have all been generated from native image features the first time they have been perceived. Afterwards, the adaption depends on whether there is new data evidence for this feature or not. If a feature has an image match and is included in the according detection, its weight is increased because α_{mat} is >1. Features that are permanently approved by data therefore are increased steadily over time. This leads to an automatic increase in hypothesis reliability that is determined by the weight of the assigned features.

When projected features are not supported by image features, the weight P_{type}^{π} is decreased because the factor α_{hyp} is <1. The reliability of a hypothesis thereby automatically decreases when no new feature evidence is available. In this case, the hypothesis is maintained by just the projected features. This inherent preservation of hypotheses even when no evidence is available, is essential to be able to track objects that are completely occluded for a short period of time. The period of time that a hypothesis is maintained in cases where no or very little image evidence is available, depends on the value of α_{hyp}. The lower this value, the faster hypothesis reliability decreases. Since these projected features are fed into the detection at every point in time, the hypothesis automatically re-strengthens when these features can be re-validated by image data after the occlusion occurred. New image features that are integrated into the detection (by voting for the same center location) also increase the reliability since they provide additional feature support for the hypothesis.

3.1.4 Automatic Generation of Object Identity Models

Besides the automatic adaption of reliability in the object detection step, the inherent inclusion of feature history results in a second advantage. By this, features that have been seen very often and thereby have a high P_{type}^{π}, also have a strong vote in the detection process. Features that have not been seen in recent history, decrease in their influence in the object detection and are removed completely after a certain time (by a threshold applied to P_{type}^{π}). This is important in cases where the visual appearance of an object changes due to viewpoint changes or environmental influences. Features that are not significant for the object any more are removed after a certain time of absence. New features which are significant for the object now, are integrated into the hypothesis automatically. By this inherent generation of an *object identity model*, we are able to reliably re-identify objects based on the standard feature codebook without the need to establish an instance specific codebook like proposed in [32]. Thus, we keep the generality of object description and simultaneously are able to re-identify single object instances. The identity models are relevant especially in cases where multiple objects occlude each other. Without the projection of hypotheses, this situation often results in indeterminable voting behavior. In practice, the strongest voting maxima is often right between the objects, since this position in the voting space gets support by features of two existing objects. In our approach, this problem is solved by the expectation projection and especially through the adaption of weights which generates the distinguishable object identity model. By matching hypothesis- with image-features before detection and consecutively adapting the weight of the resulting votes by inherently including the feature history, we can determine which image features belong to which hypotheses. Features which have been seen in a hypotheses very often are, by a high P^{π}_{type}, more likely to be assigned to this hypothesis (see Sect. 3.1.1).

3.2 Results and Evaluation

To assess the quality of our tracking compared to a feature based tracking without the projection of expectations, we consider a situation (see Fig. 8) where two people walk past each other and one person is occluded a significant part. The top row shows results of a feature based tracking with independent detection and subsequent track formation. Here, we see that at the time of people overlapping, only a single object detection is generated by the two persons. From a single image, the detection approach is not able to distinguish these two persons. The result is, that the identities are not correctly maintained. The bottom row shows the results of our tracking approach in the same situation. As we see, the object identities are preserved correctly and the approach is able to estimate the position and size of the occluded person correctly even when it is nearly completely occluded by another person.

Quantitative tracking evaluation is done with two main aspects. First, we want to show how tracking improves detection performance by stabilizing it over time. For that, we evaluate tracking in the same three image sequences we already used in Sect. 2.3.2 for standalone detection evaluation. The second aspect is the performance of tracking itself. Here, in addition to the performance of object detection, we measure how good the trajectory maintenance, namely the identity preservation is. For that, we evaluate tracking in two other image sequences which include additional difficulties for tracking.

For evaluation, we annotate every person in the video sequence with a bounding box. Since our tracking approach is in principle capable to infer the presence of occluded objects when they have been seen before (see Fig. 8), we also annotate temporary occluded persons and the occluded parts of persons if they have been "fully-visible" previously in the sequence.

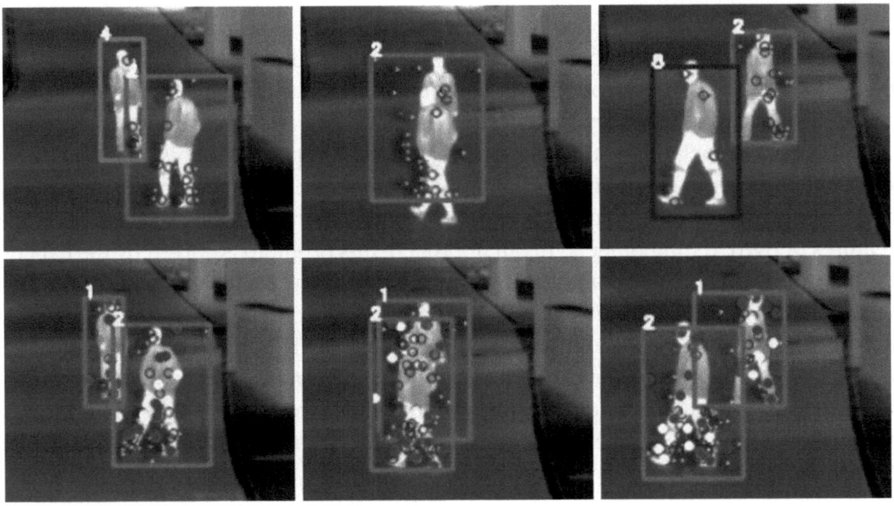

Fig. 8 Comparison of tracking results using the integration of perception and expectation (*bottom-row*) and a feature based tracking without these extensions (*top-row*). *Dots* visualize the features contributing to a person hypothesis. *Circles* in the *bottom-row* depict projected features that cannot be verified by image data but contribute to the according hypothesis

To determine whether an object hypothesis is a true- or a false positive, we use the *overlapping* criterion introduced in Sect. 2.3.2. In contrast to the evaluation of standalone detection, we only evaluate using the strongest overlap demand with a minimum 0.5 (50%) to be regarded as true positive, here. Again, only a single hypothesis is counted per ground truth object, all other hypotheses are counted as false positive for this object.

To assess tracking performance, we use the metrics:

$$\text{MOTP} = \frac{\sum_{i,t} d_t^i}{\sum_{i,t} gt_t^i} \tag{9}$$

$$\text{MOTA} = 1 - (\overline{m} + \overline{fp} + \overline{mm}) \tag{10}$$

from [4], which include a complete assessment of tracking performance, detection performance and precision.

The Multiple Object Tracking Precision (MOTP) indicates the overall exactness of detections where d_t^i is the distance (i.e., the center distance) between a true positive detection and the ground truth. Since we evaluate our tracking performance using a bounding box criterion, we do not use the distance but the overlap of detection and ground truth bounding box. Thus, MOTP in our case is the mean bounding box overlap (so 1.0 would be the best results here) of all correct detections.

The Multiple Object Tracking Accuracy (MOTA) accounts for the overall tracking performance by taking into account the miss-ratio $\overline{m}$, the false positive ratio $\overline{fp}$ and the mismatch ratio mm:

Table 1 Tracking results for sequences 1–5

Sequence	2	3	4	1	5
Frames	400	201	417	71	216
Objects (#ids)	763 (2)	1,471 (8)	1,119 (8)	301 (7)	417 (3)
MOTP	0.62	0.62	0.67	0.67	0.62
Miss rate	0.05	0.17	0.16	0.43	0.1
False positive rate	0.02	0.15	0.08	0.05	0.1
Mismatch rate	0	0.002 (4)	0.001 (2)	0	0
MOTA	*0.93*	*0.66*	*0.76*	*0.52*	*0.79*
Recall	0.95	0.82	0.84	0.57	0.9
False positive/image	0.04	1.15	0.26	0.22	0.25

$$\overline{m} = \frac{\|m\|}{\|gt\|}, \quad \overline{fp} = \frac{\|fp\|}{\|gt\|}, \quad \overline{mm} = \frac{\|mm\|}{\|gt\|} \tag{11}$$

that are accumulated over all frames ($\|gt\|$ is the number of objects in ground truth). Mismatches are counted when the object id in the tracking system changes for a ground truth object. To allow for comparison of our results with other work that only accounts for detection accuracy, we additionally show the recall (ratio of true positives and ground truth objects) and the false positives per image in the result Table 1.

The result plots for sequences 2 and 3 (compare to 3), are shown in Fig. 9. As we see, only the 50% overlap demand criterion is evaluated. The bottom curve is a plot of the standalone detection performance in this image sequence which was discussed in Sect. 2.3.2. The top curve shows detection performance when using tracking, again with a 50% bounding box overlap demand criterion. As we see from the plots, the performance increases significantly in both cases. For sequence 2, we gain a recall of 0.95 with only 0.04 false positives per image. This is immense improvement compared to the standalone detection where the highest recall was about 0.73 but with a false positive rate of nearly 1.9. This shows how strong tracking improves detection accuracy in this case. Standalone detection already had good results when using the "inside bounding criterion" or a lower overlap demand, but it was lacking the accuracy to accomplish good results with higher overlap demand. This is now accomplished in tracking. For sequence 3 the improvement is even bigger. Here we gain 0.9 recall at about 5.75 false positives per image which is an improvement of more than 0.35. Even more important, we already have a recall of 0.82 at a false positive rate of 1.15. This is an immense improvement of over 0.6 compared to the standalone detection.

These good results are confirmed by the tracking evaluation of these sequences which is shown in Table 1. Sequence 2 has a nearly perfect performance with a MOTA of 0.93 and no mismatch. This tracking performance might have been expected since, as we see in the top row of Fig. 10, there are only two persons moving around distant from each other. In the more challenging scenario in sequence 3, tracking shows a good performance too with a MOTA of 0.66. The main challenge for tracking here is

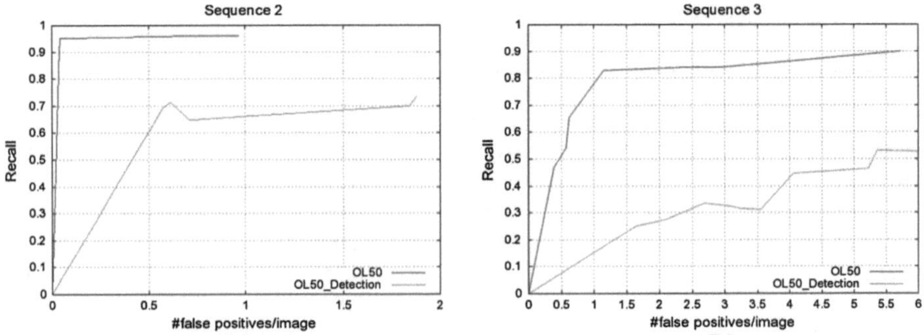

(a) **(b)**

Fig. 9 Recall/false positive curves for evaluation set 2: **a**, 3: **b**. Each chart contains two graphs which refer to the performance of tracking and standalone detection regarding the 50% bounding box overlap criterion

Fig. 10 Example tracking results of sequence 2 (*top row*) and sequence 3 (*bottom row*)

identity maintenance for the four persons in the back of the scene which appear at a very low scale. The small appearance might lead to short term failures of the detection, even if improved by tracking. These breaks in tracks together with the moving camera can lead to mismatches. This happens when a person is re-detected after a short failure, but the position has changed significantly due to camera motion. In this case the tracks cannot be associated with each other and a new object hypothesis with a new ID is instantiated. We can see this in the sample results in the bottom row of Fig. 10 and from the mismatch rate which counts four mismatches in this sequence.

To analyze the tracking in more depth, we evaluated it in another sequence (4). This sequence is more challenging for tracking because besides the moving camera, people are moving around a lot more than in sequence 2 (where people mainly

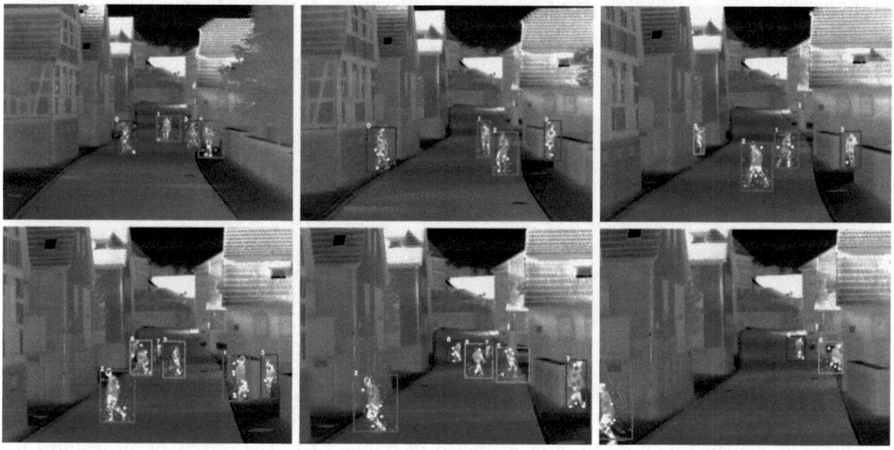

Fig. 11 Example tracking results of sequence 4

moved towards the camera) which leads to a lot of occlusions between people. This is particular difficult for tracking in infrared, because, specifically when the camera is moving, there is not much information that can be used for identity maintenance in these situations. Totally, there are eight different persons in this sequence, four of which are running from one side to another in the background of the scene, thus appearing very small, which is another difficulty here. Sample results of this sequence are shown in Fig. 11. From these and the results in Table 1, we can see that the tracking performs well, with a MOTA of 0.76 and only two mismatches, even under these difficult circumstances, where three problems, namely a moving camera, people appearing at both, very low and high scales, and people occluding each other, coincide.

Sequence 1 comes with a lot of difficulties considering person tracking. For our tracking strategy, specifically the strong camera motion is a problem since we propagate expectations based on a dynamics modelling using a Kalman filter. This Kalman filter is appropriate in cases of static cameras because people do not move that much from frame to frame. Even in cases of slight motion like in sequences 3 and 4, this dynamics model proved to be sufficient. In sequence 1, camera motion is very strong, which leads to strong shifts in the image position of persons. This makes tracking a challenging problem, specifically in infrared where the appearance of person is very alike and position is an important cue. This position shift leads to accuracy problems in detection (when using the propagation strategy) and when looking at the tracking (and not the detection) performance, to identity changes of single persons and between objects. Since this is as important as the detection performance, we introduce motion compensation model that copes with strong camera motion and allows for tracking in these situations.

Fig. 12 Calculation of shift vectors between frames. *Top row* shows tracking results for time T. *Bottom row* visualizes calculation of shift vectors for the next frame: each feature of a person hypothesis is matched with all features extracted in frame $T+1$. The offsets of all features the similarity of which is high enough are recorded with their weights to calculate the overall motion between frames

3.3 Tracking Under Strong Camera Motion

As mentioned in the last section, the dynamics model using a Kalman filter is not sufficient to track person from a strongly moving camera. In this section, we introduce a method that makes tracking completely independent from camera motion without explicitly calculating a motion compensation, e.g., with a homography. Our approach fits into the detection and tracking strategy and thus does not have to employ any other methods. We replace the Kalman prediction component of the system by a calculation of shift vectors between frames. For that, as shown in Fig. 12, each feature in every object model (hypothesis) is matched with the image features of the next frame. For every feature–feature combination the similarity of which is high enough, the shift vector from the last to the current frame (the movement of the person) and its according weight, which is determined by feature similarity (see Eq. 1), is recorded. As we see in Fig. 12 for some example features, this is done for all hypotheses. The shift vectors are then transferred to a 2D voting space where each vector votes for a certain offset (see Fig. 13a). As we see in Fig. 13b, these votes have different weights assigned (visualized by the size of the circle) and were generated by different hypotheses (indicated by different colors). We can see here, that the shift that refers to the camera motion should build a cluster in this space. Indeed this cluster must not necessarily be very dense since people might walk into different directions which distorts the maximum since it dilutes the motion.

In the next step, as shown in Fig. 13c, a maximum search in this space is conducted. For that we use mean shift, since this allows for accounting for imprecision by increasing the kernel size and thus is capable to cope with the impreciseness generated by possible ego motion of people. This technique is preferable over e.g., calculating a homography and registering the frames for two reasons. First it fits into our strategy

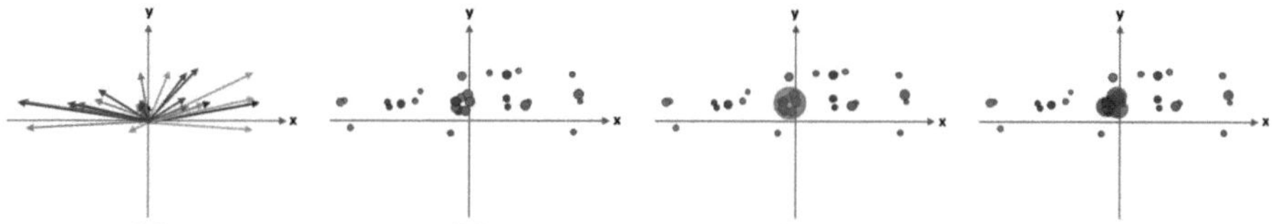

Fig. 13 Motion compensation for person tracking: **a** transfer of feature offsets (see Fig. 12) to 2D voting space. **b** Votes with assigned weight for motion between frames. **c** Global maximum mean shift search in the voting space to determine camera motion between frames. **d** Mean shift search to determine people motion between two frames

and directly delivers the assignment we need for feature propagation. Second we cannot expect to be able to calculate an exact homography since people might move in different directions which would distort an exact registration. In our approach this is absorbed by the two stage strategy and the imprecision that is deliberately tolerated in the first stage where only the global motion matters.

Now that we know the approximate camera motion, we can account for the ego motion of persons. This is done in the next stage (Fig. 13d), where another mean shift search is applied, now for every hypothesis independently. Starting from the position of the global maximum, the hypothesis specific mean shift searches for the shift position of a certain hypothesis using only the votes of this specific hypothesis. The choice of the global maximum as a starting point is necessary because if the shift of each hypothesis is calculated independently, specifically in infrared where people look much alike, this might lead to permutations between objects.

This overall strategy ensures correct identity maintenance (correct assignment over time) in two ways. Under the assumption that the spatial collocation of people in the scene stays the same, which means that people do not change relative positions, the number of feature shift vectors that constitutes the correct sensor motion (image content displacement) clearly should be bigger than the number of shift vectors that are constituted by incorrect assignments. The second assumption is that the feature similarity between the same objects at two points in time, is higher than the similarity between different objects. In this case, the correct shift vectors are weighted higher than those corresponding to wrong assignments. Another important point regarding similarity on feature level is that even when objects are from the same class (here persons), the feature extraction responds to object specifics, which means that a certain object can contain features that are only found on this particular object and thus have no match at all on other objects of the same class. Using this combination of spatial consistency and appearance information to calculate the overall shift vector, we gain a high stability under strong motion even if one of the assumptions does not hold in some cases.

We can see this in the tracking evaluation of sequences 1 and 5. For sequence 1 we already showed the standalone detection results in Sect. 2.3.2. The comparison to tracking in Fig. 14 shows that detection performance only increases slightly. That is because some people are not detected at all. Since tracking is only able to stabilize

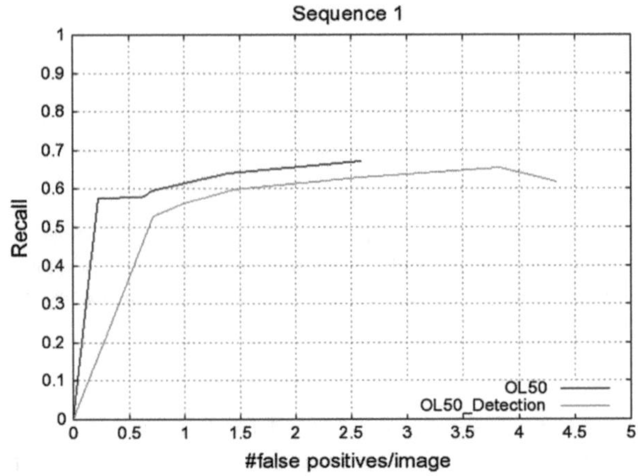

Fig. 14 Recall/false positive curves for sequence 1. The chart contains two graphs that refer to the performance of tracking and standalone detection regarding the 50% bounding box overlap criterion. Here, we see a slight improvement of detection performance using tracking. Performance increase is rather minor because in this sequence some persons are not detected at all. Since tracking requires a person to be detected once, it cannot increase performance significantly in this case

Fig. 15 Example tracking results of sequence 1

Fig. 16 Example tracking results of sequence 5

detection after an object has been detected once, but not to improve initial detection performance, this cannot be solved by tracking. Tracking performance itself, as shown in Table 1, is good with no mismatches. Sample results of this sequence are shown in Fig. 15. To show the usability of our new dynamics model, we evaluate it in another sequence (5). Here, three persons are to be tracked from a camera installed on a vehicle which drives on a vehicle test track. The test track has artificial ground waves which lead to very strong motion of the vehicle (and thus of the camera). Sample results for this sequence are shown in Fig. 16. The evaluation results in Table 1 show that no mismatch happened in this sequence. This shows that the tracking strategy works well even with a strongly moving camera.

4 Conclusion

In this chapter, we presented a generic approach to detect and track persons based solely on local (SURF) image features. We showed that this approach is specifically suited to detect and track persons in infrared image sequences and is capable of tracking people consistently in case of strong camera movement. For that, we introduced a novel tracking technique, that combines expectation gathered in the past with newly available data on the level of features. By integrating this tracking technique directly into the ISM based detection approach, we gain an improvement in detection performance itself. This was shown in the evaluation of standalone detection in three different image sequences with different challenges for person detection. We showed that this detector performs well in detecting persons per se but is rather inaccurate in terms of bounding boxes. The evaluation of tracking in these sequences showed the immense detection improvement by the stabilizing effect of tracking which compensates short time detection failures and increases detection accuracy significantly. Furthermore, we showed that the integrated tracking approach performs well regarding trajectory correctness (identity preservation) and is able to track persons through short term occlusions even in infrared image sequences where tracking is more difficult due to the very similar appearance of persons. In addition we showed that the approach copes with camera motion and introduced a novel approach to model motion dynamics during tracking that is able to track persons even under difficult conditions with strong camera motion. In addition to just detect and track people as compound, we showed how this local feature approach can be used to classify a person's body parts which can be input to a higher level scene interpretation.

References

1. Andriluka, M., Roth, S., Schiele, B.: People-tracking-by-detection and people-detection-by-tracking. In: Proceedings of the Conference on Computer Vision and Pattern Recognition, pp. 1–8 (2008)
2. Bay, H., Ess, A., Tuytelaars, T., Van Gool, L.: Speeded-up robust features. Comput. Vis. Image Underst. **110**(3), 346–359 (2008)
3. Belongie, S., Malik, J., Puzicha, J.: Shape matching and object recognition using shape contexts. Trans. Pattern Anal. Mach. Intell. **24**(4), 509–522 (2002)
4. Bernardi, K., Elbs, A., Stiefelhagen, R.: Multiple object tracking performance metrics and evaluation in a smart room environment. In: The IEEE International Workshop on Visual Surveillance, pp. 219–223 (2006)
5. Berrabah, S.A., De Cubber, G., Enescu, V., Sahli, H.: MRF-based foreground detection in image sequences from a moving camera. In: Proceedings of the IEEE International Conference on Image Processing, pp. 1125–1128 (2006)
6. Bertozzi, M., Broggi, A., Grisleri, P., Graf, T., Meinecke, M.: Pedestrian detection in infrared images. In: Proceedings of the IEEE Intelligent Vehicles Symposium, pp. 662–667 (2003)
7. Conaire, C.O., Cooke, E., O'Connor, N., Murphy, N., Smearson, A.: Background modeling in infrared and visible spectrum video for people tracking. In: Proceedings of the IEEE Conference on Computer Vision and Pattern Recognition, pp. 20–25 (2005)
8. Dalal, N., Triggs, B.: Histograms of oriented gradients for human detection. In: Proceedings of the IEEE Conference on Computer Vision and Pattern Recognition, pp. 886–893 (2005)
9. Davis, J., Sharma, V.: Background-subtraction using contour-based fusion of thermal and visible imagery. Comput. Vis. Image Underst. **106**(2–3), 162–182 (2007)
10. Davis, J.W., Sharma, V.: Robust background-subtraction for person detection in thermal imagery. In: Proceedings of the Conference on Computer Vision and Pattern Recognition Workshop, pp. 128 (2004)
11. Enzweiler, M., Gavrila, D.M.: Monocular pedestrian detection: survey and experiments. IEEE Trans. Pattern Anal. Mach. Intell. **31**(12):2179–2195 (2009)
12. Fang, Y., Yamada, K., Ninomiya, Y., Horn, B., Masaki, I.: Comparison between infrared-image-based and visible-image-based approaches for pedestrian detection. In: Proceedings of the IEEE Intelligent Vehicles Symposium, pp. 505–510 (2003)
13. Felzenszwalb, P., Mcallester, D., Ramanan, D.: A discriminatively trained, multiscale, deformable part model. In: IEEE Conference on Computer Vision and Pattern Recognition (2008)
14. Gammeter, S., Ess, A., Jaeggli, T., Schindler, K., Leibe, B., Van Gool, L.: Articulated multi-body tracking under egomotion. In: Proceedings of the European Conference Computer Vision, pp. 816–830 (2008)

15. Gavrila, D., Munder, S.: Multi-cue pedestrian detection and tracking from a moving vehicle. Int. J. Comput. Vis. **73**(1), 41–59 (2007)
16. Haritaoglu, I., Harwood, D., Davis, L.: W4s: a real-time system for detecting and tracking people in 2.5 d. In: Proceedings of the European Conference on Computer Vision, pp. 877+ (1998)
17. Harris, C., Stephens, M.: A combined corner and edge detection. In: Proceedings of the Alvey Vision Conference, pp. 147–151 (1988)
18. Hu, W., Tan, T., Wang, L., Maybank, S.: A survey on visual surveillance of object motion and behaviors. IEEE Trans. Syst. Man Cybern. **34**(3), 334–352 (2004)
19. Jaccard, P.: Nouvelles recherches sur la distribution florale. Bull. Soc. Vaudoise Sci. Naturelles **4**(3), 223–370 (1908)
20. Jüngling, K., Arens, M.: Detection and tracking of objects with direct integration of perception and expectation. In: Proceedings of the International Conference on Computer Vision, ICCV Workshops, pp. 1129–1136 (2009)
21. Jüngling, K., Arens, M.: Feature based person detection beyond the visible spectrum. In: Proceedings of the IEEE Computer Society Conference on Computer Vision and Pattern Recognition, CVPR Workshops, pp. 30–37 (2009)
22. Klinger, V., Arens, M.: Ragdolls in action–action recognition by 3d pose recovery from monocular video. In: Proceedings of the IADIS International Conference on Computer Graphics, Visualization, Computer Vision and Image Processing, pp. 219–223 (2009)
23. Leibe, B., Leonardis, A., Schiele, B.: Combined object categorization and segmentation with an implicit shape model. In: ECCV Workshop on Statistical Learning in Computer Vision, pp. 17–32 (2004)
24. Leibe, B., Leonardis, A., Schiele, B.: Robust object detection with interleaved categorization and segmentation. Int. J. Comput. Vis. **77**(1–3), 259–289 (2008)
25. Leibe, B., Schindler, K., Cornelis, N., Van Gool, L.: Coupled object detection and tracking from static cameras and moving vehicles. IEEE Trans. Pattern Anal. Mach. Intell. **30**(10), 1683–1698 (2008)
26. Leibe, B., Schindler, K., Van Gool, L.: Coupled detection and trajectory estimation for multi-object tracking. In: Proceedings of the International Conference on Computer Vision, pp. 1–8 (2007)
27. Leykin, A., Hammoud, R.: Robust multi-pedestrian tracking in thermal-visible surveillance videos. In: Proceedings of the Conference on Computer Vision and Pattern Recognition Workshop, pp. 136+ (2006)
28. David Lowe, G.: Distinctive image features from scale-invariant keypoints. Int. J. Comput. Vis. **60**(2), 91–110 (2004)
29. Munkres, J.: Algorithms for the assignment and transportation problems. J. Soc. Industrial Appl. Math. **5**, 32–38 (1957)
30. Nanda, H., Davis, L.: Probabilistic template based pedestrian detection in infrared videos. In: Proceedings of the IEEE Intelligent Vehicle Symposium, vol. 1, pp. 15–20 (2002)
31. Ren, Y., Chua, C., Ho, Y.: Statistical background modeling for non-stationary camera. Pattern Recognition Lett. **24**(1–3), 183–196 (2003)
32. Seemann, E., Fritz, M., Schiele, B.: Towards robust pedestrian detection in crowded image sequences. In: Proceedings of the IEEE Conference on Computer Vision and Pattern Recognition, pp. 1–8 (2007)
33. Stauffer, C., Grimson, W.: Adaptive background mixture models for real-time tracking. In: Proceedings of the IEEE Conference on Computer Vision and Pattern Recognition, pp. 246–252 (1999)
34. Suard, F., Rakotomamonjy, A., Bensrhair, A., Broggi, A.: Pedestrian detection using infrared images and histograms of oriented gradients. In: Proceedings of the IEEE Intelligent Vehicles Symposium, pp. 206–212 (2006)
35. Tuytelaars, T., Mikolajczyk, K.: Local Invariant Feature Detectors: A Survey. Now Publishers Inc., Hanover (2008)

36. Viola, P., Jones, M.: Rapid object detection using a boosted cascade of simple features. In: Proceedings of the IEEE Conference on Computer Vision and Pattern Recognition, vol. 1, pp. 511–518 (2001)
37. Wren, C., Azarbayejani, A., Darrell, T., Pentland, A.: Pfinder: real-time tracking of the human body. In: Proceedings of the International Conference on Automatic Face and Gesture Recognition, pp. 51–56 (1996)
38. Wu, B., Nevatia, R.: Detection and tracking of multiple, partially occluded humans by bayesian combination of edgelet based part detectors. Int. J. Comput. Vis. **75**(2), 247–266 (2007)
39. Xu, F., Fujimura, K.: Pedestrian detection and tracking with night vision. In: Proceedings of the IEEE Intelligent Vehicle Symposium, pp. 21–30 (2002)
40. Yilmaz, A., Javed, O., Shah, M.: Object tracking: a survey. ACM Comput. Surv. **38**(4), 13+ (2006)

Appearance Learning for Infrared Tracking with Occlusion Handling

Guoliang Fan, Vijay Venkataraman, Xin Fan and Joseph P. Havlicek

Abstract This chapter discusses the issue of appearance learning for infrared target tracking with occlusion handling. The problem is cast in a co-inference framework, where both adaptive Kalman filtering (AKF) and particle filtering are integrated together to learn target appearance and to estimate target kinematics in a sequential manner. We propose a dual foreground–background appearance model that incorporates the pixel statistics in both foreground and background areas for an effective target representation. Appearance learning is formulated as an AKF problem that can be approached by either covariance or correlation methods for noise estimation. Moreover, occlusions can be easily detected by analyzing the Kalman filtering residuals. Experiments on real infrared imagery show that correlation-based AKF outperforms the covariance-based one as well as traditional histogram similarity-based approaches with near sub-pixel tracking accuracy and robust occlusion handling.

Keywords Appearance learning · Histogram filtering · Target tracking · Kalman filters · Infrared Tracking · Occlusion handling · FLIR

G. Fan (✉) · V. Venkataraman
School of Electrical and Computer Engineering, Oklahoma State University,
Stillwater OK, USA
e-mail: guoliang.fan@okstate.edu

V. Venkataraman
e-mail: vvenka@okstate.edu

X. Fan
School of Software, Dalian University of Technology, Dalian, China
e-mail: xin.fan@ieee.org

J. P. Havlicek
School of Electrical and Computer Engineering, University of Oklahoma,
Norman OK, USA
e-mail: joebob@ou.edu

R. Hammoud et al. (eds.), *Machine Vision Beyond Visible Spectrum*,
Augmented Vision and Reality, 1, DOI: 10.1007/978-3-642-11568-4_2,

1 Introduction

Infrared target tracking has remained a challenging problem in the field of image processing and understanding. Infrared imagery acquired under actual field condition is typically characterized by strong structured background clutter, poor SNR, and significant ego-motion of the sensor relative to the target. In addition to these difficulties arising from sensor and environmental factors, the targets of interest are also highly maneuverable. Therefore, their observed signatures may exhibit profound non-stationary variations over relatively short time scales making it difficult to maintain a robust track lock over long time scales. This phenomenon of the target representation deviating from its true signature due to accumulated tracking errors has been referred to variously as the "drifting problem" in [11, 34], the "template update problem" in [12, 17, 22, 36], and a "stale template condition" in [14]. These are challenges that are exemplified by the well-known AMCOM[1] closure sequences [7, 9, 15, 27, 43, 47, 50] and the VIVID dataset,[2] which will be used as illustrative examples in this work.

An example of this non-stationary variation is shown in Fig. 1a. Here, a longwave imaging sensor is situated on an airborne platform that closes on a pair of maneuvering ground vehicles. Profound changes in the target's appearance are observed between frames 24 and 165 over a time scale of only a few seconds and arise primarily from the relative motion between the sensor and the target. There is substantial magnification that results from the sensor closing on the target and pose change that results from the target executing an aggressive turning maneuver. While the second vehicle in Fig. 1a exhibits a strong signature, the lead vehicle is much dimmer and is barely visible amid the surrounding clutter, demonstrating that brightness alone cannot be used as the sole basis for reliable detection and tracking. Rather, more sophisticated techniques are generally required for representing the target appearance and for adapting to (e.g., learning) complex appearance changes that occur over time. Further, in many cases the target being tracked may move out of the sensor's view or become occluded thereby significantly altering the observed appearance. An example is shown in Fig. 1b where the target being observed moves behind a tree along its path and thereby disappears from the sensor's view. While it is important to adapt the appearance model to accommodate variations in the target signature it is equally important to avoid learning the appearance of occluding objects or the background. A robust tracker must be able to quickly adapt to profound variations in the target signature and must suspend adaptations when the target of interest is occluded or has left the scene.

In this work we address the issues of appearance learning and robust tracking in an unified *co-inference* framework similar to the one proposed in [45]. Histogram-based appearance learning is explicitly combined with the estimation of the target position and size, achieving sub-pixel accuracy. Specifically, we utilize a

[1] Available from the Center for Imaging Science at the Johns Hopkins University (http://cis.jhu.edu).

[2] https://www.sdms.afrl.af.mil/main.php

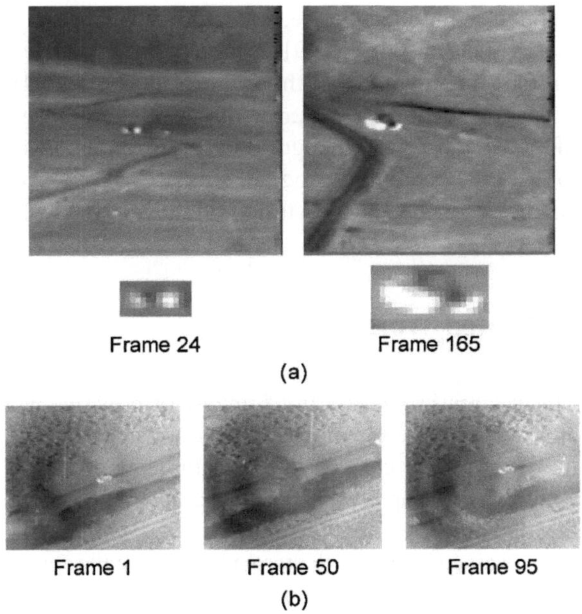

Fig. 1 **a** Example of nonstationary target signature evolution in AMCOM LWIR run *rng18_17*. There are two vehicles where the lead one is barely visible, and the second one (the one of interest) is clearly visible. *Top row*: observed frames. *Bottom row*: closeup views of the target. **b** Example of a target being occluded by foliage in the VIVID dataset

dual foreground–background appearance representation that involves a total of four histograms, including histograms of the pixel intensities and of the local standard deviations computed over both the target region and the immediately surrounding background. The intensity histograms for both the target and background are estimated in each frame along with the target position and size. This coupled estimation forms the *co-inference* process, in which the inference of histograms relies on that of target kinematics, and vice versa. The estimation of the histograms, i.e., appearance learning, is achieved by a bank of adaptive Kalman filters (AKFs), where the unknown process and measurement noise variances are estimated simultaneously using the recently developed autocovariance least squares (ALS) method [32, 33]. The main tracker is a particle filter where the state vector gives the target position and size. The likelihood function depends on the adaptive appearance model. Further, we devise a track loss detection scheme embedded into the appearance learning process. Aside from the accurate estimation of the target appearance, temporary track losses can also be detected by examining the Kalman filter residuals, byproducts of AKFs at each step. Hence, we are able to tackle both the "drifting problem" and "temporary track losses" in this *co-inference* framework.

The main contributions of this chapter include the application of the ALS-based AKF method in visual target tracking for the first time, the development of a robust appearance learning algorithm based on a quad of dual foreground–background histograms, devise a track loss detection scheme embedded into the appearance learning process, and the integration of these techniques in a co-inference framework to achieve sub-pixel tracking accuracy. The new tracking and appearance learning techniques introduced here, which are developed in Sects. 4 and 5, are distinct from those given in [34, 35, 49] in the use of a particle filter as opposed to the mean-shift algorithm and from those given in [29, 30, 34] in the use of histogram-based appearance learning. Experimental results are given in Sect. 7 where performance of the ALS-based AKF algorithm for histogram learning is quantitatively compared to that obtained with histogram similarity based and covariance matching methods and track loss detection is tested on the VIVID dataset.

2 Related Work

This section provides a brief review of previous works on the important issues of appearance representation, learning and occlusion handling. Target representations may be broadly categorized as parametric, where a statistical model is typically assumed that captures the key characteristics of the target appearance in a way that facilitates estimation of the model parameters continuously online [13], or non-parametric, where the target appearance is characterized by empirically derived features that can be updated online during tracking [2, 6]. Such features may include kernel-based windows [8, 39, 44], nonparametric or semiparametric contours [48], templates [8], shape descriptors [39], or local statistics [8, 49] including intensity histograms and their moments. Histogram-based features have been widely adopted as the appearance model in many recent trackers [4, 6, 35, 49, 52], due to their simplicity and their scale and rotation invariance properties [6, 21, 49].

Significant efforts have been directed towards developing methods for online appearance learning in order to combat the "drifting problem" [11, 13, 22, 23, 29]. In the case of parametric models, a sophisticated model combining stable, wandering, and outlier components in a Gaussian mixture model (GMM) was proposed in [13], where the model was updated via an expectation maximization (EM) algorithm. GMM based appearance learning was also applied in [10], where a mean-shift algorithm was used to update the parameters online. These methods rely on elaborate parametric models and are effective for tracking extended targets with large spatial signatures. However, for small targets such as those shown in Fig. 2, there may not be enough pixels on the target to achieve robust and statistically significant parameter estimation. In the context of appearance update schemes for non-parametric features, drift correction strategies for template tracking were proposed in [16, 22]. For histogram-based target representations, appearance learning is generally accomplished by iteratively updating a reference histogram [18, 35, 51]. Typically, the new reference histogram at each iteration is given by a linear weighting of the previous reference histogram and the most recent observation, where the weighting coefficient may be based on an appropriate measure of histogram similarity. While such techniques are often effective for adapting the appearance model when the target has a large spatial extent, they can be susceptible to drifting problems, particularly when applied to smaller targets.

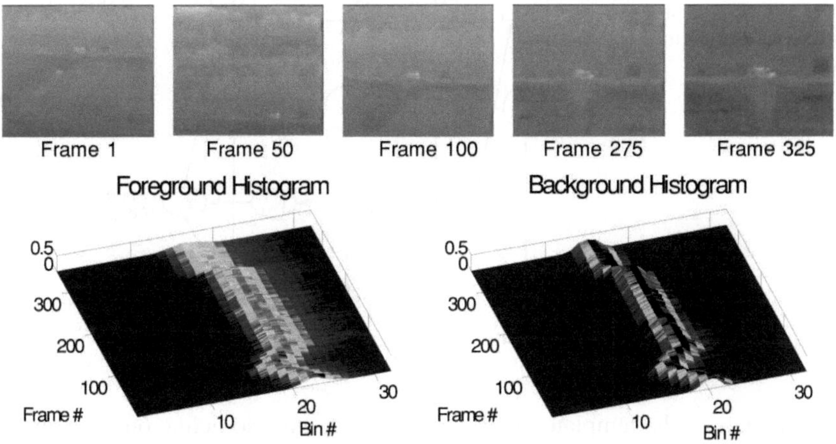

Fig. 2 The variations of the target appearance in the sequence LW-17-01. Some sample infrared frames are shown above, and intensity histograms of foreground and background below

Time traces of the normalized pixel intensity histograms for the target and local background in AMCOM LWIR sequence *rng17_01* are given in Fig. 2 along with several raw video frames. In the early part of the sequence the target is relatively dim and is barely distinguishable from the background. There is considerable overlap between the target and background histograms throughout, as is typical for sequences acquired under practical field conditions of this nature. Accurate estimation of the histograms is critical in such cases, since the accumulation of small errors over time can corrupt the target model and ultimately cause the track filter to lock onto the background structure and fail.

Improved histogram estimation was achieved by modeling the temporal evolution of the reference histogram in an adaptive Kalman filtering (AKF) framework in [35]. In [30, 34], the AKF measurement noise variance was estimated from the first frame and was assumed stationary, while the process noise variance was estimated online using covariance matching [25]. A robust Kalman filter was also developed for appearance learning in [29], where the process noise was assumed known and covariance matching was used to estimate the variance of the innovations.

Many other works also address the issues of temporary track losses and occlusions. Typically, occlusion can be detected by investigating the distance between candidates and reference representations. The distance between the contours of objects was used in [48]. Latecki and Miezianko incorporated motion cue into the definition of the template distance [17]. Wu et al. explicitly introduced a state variable as the indicator for occlusion into the dynamic Bayesian networks in order to estimate the probability of occlusion. In their approach, the likelihood is also defined based on the template distances [46]. For histogram representations, the percentage of outliers are taken as an index to detect occlusion in [30], and the outliers are classified based on Kalman

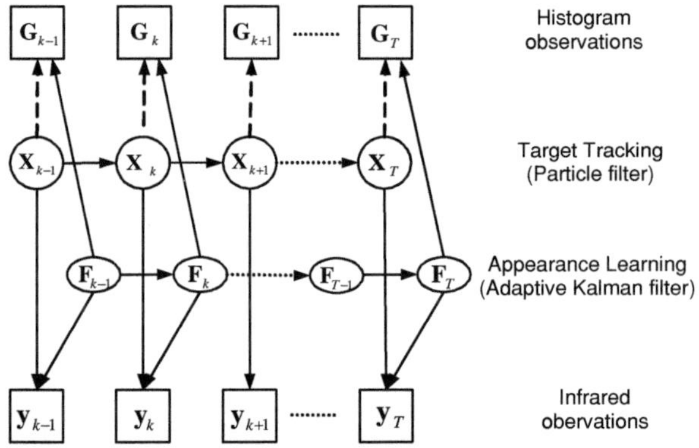

Fig. 3 Appearance learning for target tracking in a co-inference paradigm

filter residuals of the template pixels. In this work we also define our indicator for occlusion using the residuals obtained by AKFs, which works well for short period of track loss and occlusions.

3 Problem Formulation

In contrast to most traditional appearance updating schemes, we explicitly model the evolution and observation processes of the target's appearance histograms. Thus, appearance learning, i.e., the update of target's appearance histograms, is formulated as a sequential state estimation problem. In this section, we introduce a graphical model, as shown in Fig. 3, that integrates estimation of the target's appearance (histogram bins) $\mathbf{F}_{1:T}$ with that of the target's kinematic states, $\mathbf{X}_{1:T}$. The kinematics $\mathbf{X}_{1:T}$ and appearance $\mathbf{F}_{1:T}$ are coupled by means of their respective observations, $\mathbf{y}_{1:T}$ and $\mathbf{G}_{1:T}$. At any time instant k the appearance $\mathbf{F}_{k-1}$ determines the likelihood of finding a target in an arbitrary area within the observed image and a tracking gate given by $\mathbf{X}_k$ is determined based on this likelihood. Then the appearance histograms ($\mathbf{G}_k$) determined from within the tracking gate act as observations to update the target appearance from $\mathbf{F}_{k-1}$ to $\mathbf{F}_k$. Wu and Huang show that the estimation of hidden states interacting with common observations invokes a *co-inference* process, denoted as:

$$\begin{aligned} P_{X_k} &\approx \mathcal{X}(\mathbf{Z}_k, E[\mathbf{F}_k|\mathbf{Z}_k]), \\ P_{F_k} &\approx \mathcal{F}(\mathbf{Z}_k, E[\mathbf{X}_k|\mathbf{Z}_k]), \end{aligned} \tag{1}$$

where $\mathbf{Z}_k$ includes the observations of kinematics and appearance histograms, P_{X_k} and P_{F_k} are the probability distributions of $\mathbf{X}_k$ and $\mathbf{F}_k$, and $\mathcal{X}(\cdot)$ and $\mathcal{F}(\cdot)$ represent the inference processes for these two distributions, respectively. $E(\cdot)$ denotes mathematical expectation, which can be approximated by a statistical estimate in practice. Equation 1 suggests that the probability of $\mathbf{X}_k$ relies on the expectation of $\mathbf{F}_k$, and the expectation of $\mathbf{X}_k$ is used to calculate the distribution of $\mathbf{F}_k$.

We take advantage of this co-inference problem as a joint estimation of target kinematic and appearance states, and are able to apply two separate inference algorithms with the estimate of one hidden state involved in the other. Recursive Bayesian filters are well suited for inference when the dynamic evolution and observation processes for $\mathbf{X}_k$ and $\mathbf{F}_k$ are explicitly modeled. The Bayesian filtering steps for kinematics estimation at time step k, $p(\mathbf{X}_k|\mathbf{y}_k)$, from its previous distribution $p(\mathbf{X}_{k-1}|\mathbf{y}_{k-1})$ is given by [1]:

$$p(\mathbf{X}_k|\mathbf{y}_{k-1}) = \int_{\mathbf{X}_{k-1}} p(\mathbf{X}_k|\mathbf{X}_{k-1})p(\mathbf{X}_{k-1}|\mathbf{y}_{k-1})\, d\mathbf{X}_{k-1}, \tag{2}$$

$$p(\mathbf{X}_k|\mathbf{y}_k) \approx p(\mathbf{y}_k|\mathbf{X}_k, \bar{\mathbf{F}}_{k-1})p(\mathbf{X}_k|\mathbf{y}_{k-1}). \tag{3}$$

It should be noted that the estimate of appearance, $\bar{\mathbf{F}}_{k-1}$, is embedded in the inference of $\mathbf{X}_k$. Analogously, the inference of histograms can also be obtained by recursive Bayesian filtering with the estimate of $\mathbf{X}_k$ involved. These filters are numerically implemented by particle filtering and adaptive Kalman filters due to different physical characteristics and mathematical assumptions on the dynamic processes of target kinematics and appearance, respectively.

Due to the strong ego motion and maneuvering actions present in typical infrared imagery, the kinematic transition, i.e., $p(\mathbf{X}_k|\mathbf{X}_{k-1})$ in (2), can hardly be characterized by linear equations. Moreover, the calculation of histogram observations of a target in (3) requires non-linear operations from image pixels in the region given by $\mathbf{X}_k$. Hence, a particle filtering based technique is necessary for the estimation of $\mathbf{X}_k$ by (2) and (3), which is detailed in Sect. 5. Here the estimate of appearance histograms is used for likelihood evaluation, $p(\mathbf{y}_k|\mathbf{X}_k, \bar{\mathbf{F}}_{k-1})$ rather than the proposal density that generates samples of $p(\mathbf{X}_k)$ in [45].

On the other hand, the dynamics of appearance histograms, $p(\mathbf{F}_k|\mathbf{F}_{k-1})$, can be reasonably approximated by a Gaussian noise driven process that reflects the actual appearance variation in infrared imagery. Therefore, a Kalman filter becomes suitable tool for solving the problem of histogram based appearance learning. Interestingly, it is possible to deal with abrupt appearance variations due to occlusion through a byproduct of the Kalman estimation, which is elaborated in Sect. 6. The appearance histograms obtained from the image pixels within the region given by the estimate of kinematics, $\bar{\mathbf{X}}_k$, can be regarded as a direct observation of the true state of appearance histograms corrupted by Gaussian noise. Therefore the estimate of $\mathbf{X}_k$ is naturally brought into the measurement based update step in the Kalman filtering process in order to achieve *co-inference*. However, the robustness and generalization of the estimation becomes questionable if the noise parameters of the filter are set in an ad hoc method, because targets may present significant appearance variations when captured in different physical conditions (e.g., weather and temperature) even by IR sensors with identical specifications. The adaptive estimation of these noise parameters turns out to be the key to the success of this AKF-based appearance learning. Two specific AKFs will be discussed that are compared with the traditional histogram similarity method.

4 Histogram-Based Appearance Learning

Let $\mathbf{y}_k$ be the k th infrared frame, which is assumed to contain a single target region of potential interest. Let $\mathbf{g}_k = \{g_k^b\}_{b=1,\dots,N_b}$ be the observed normalized histogram of the target computed from frame $\mathbf{y}_k$, where $\sum_{b=1}^{N_b} g_k^b = 1$ with the histogram being discretized into N_b bins. Similarly, let $\mathbf{f}_k = \{f_k^b\}_{b=1,\dots,N_b}$ be the reference histogram, which provides an idealized model of the object appearance at time k. Our objective is to estimate the present appearance model $\mathbf{f}_k$ by incorporating the current tracker observation $\mathbf{g}_k$ into the previous appearance model $\mathbf{f}_{k-1}$. Histogram-based appearance learning can be formulated as a time-varying linear filtering, defined as:

$$\mathbf{f}_k = \boldsymbol{\xi}_k \odot \mathbf{g}_k + (\mathbf{1} - \boldsymbol{\xi}_k) \odot \mathbf{f}_{k-1}, \tag{4}$$

where "$\odot$" represents the Hadamard (or Schur) product and $\mathbf{1}$ is vector with all entries equal to one. $\boldsymbol{\xi}_k = \{\xi_k^b\}_{b=1,\dots,N_b}$ controls the balance between the previous reference model $\mathbf{f}_{k-1}$ and the present tracker observation $\mathbf{g}_k$, $0 \le \xi_k^b \le 1$ is the time dependent filter coefficient of the b th bin. Continuous modulation of $\boldsymbol{\xi}_\mathbf{k}$ to comply with appearance variations is key for effective appearance learning.

The remainder of this section discusses three different learning methods that share the time-varying linear filtering form defined in (4) and only differ in the computation of $\boldsymbol{\xi}_\mathbf{k}$. First, we discuss the method using histogram similarity HS where all bins are updated with the same weight ($\xi_k^b = \xi_k$, $b = 1, \dots, N_b$). After briefly reviewing the basic Kalman filter, we introduce two AKF methods that use different methods to estimate the system noise parameters.

4.1 Histogram Similarity

In the popular HS method, the weighing coefficient is computed by measuring the histogram similarity [18]. One common metric is the Bhattacharyya distance [6]. In practice, the histogram intersection metric [41] was found to be more suitable for FLIR imagery and is defined as

$$\mathrm{d}(\mathbf{f}_{k-1}, \mathbf{g}_k) = \sum_{i=1}^{N_b} \min(f_{k-1}^i, g_k^i), \tag{5}$$

Since histogram similarity is defined using information from all bins, in this method each bin is updated using the same weight ξ_k defined as

$$\xi_k = 1 - \mathrm{d}(\mathbf{f}_{k-1}, \mathbf{g}_k). \tag{6}$$

When the observed ($\mathbf{g}_k$) and reference histograms ($\mathbf{f}_{k-1}$) are very similar (i.e., small ξ_k), very little information from the observed histogram is incorporated in the learning process. On the other hand, when there is a sudden change in the target appearance and the two histograms become dissimilar (i.e., large ξ_k), the observed histogram is more influential in the learning process. However, rapid adaptation of the observed histogram and discarding the historical information contained in $\mathbf{f}_{k-1}$ can be potentially problematic if the tracker is distracted by background clutter or if the image is corrupted by noise.

4.2 Kalman Filtering

To formulate appearance learning in the context of Kalman filtering, we define the system and observation models for each bin b of the histogram as

$$f_k^b = f_{k-1}^b + w_{k-1}^b, \tag{7}$$

$$g_k^b = f_k^b + v_k^b, \tag{8}$$

where w_{k-1}^b and v_k^b are the process and observation noises that are assumed to be zero-mean iid Gaussian with variances $\sigma_{wb}^2(k)$ and $\sigma_{vb}^2(k)$, respectively. The model described in (7) and (8) hypothesizes that the histogram evolution and observations are affected only by white noise. The state prediction and update equations are given by:

$$\text{State prediction:} \quad \hat{f}_{k|k-1}^b = \hat{f}_{k-1}^b \tag{9}$$

$$\text{Covariance prediction:} \quad p_{k|k-1}^b = p_{k-1}^b + \sigma_{wb}^2(k-1) \tag{10}$$

$$\text{Kalman gain:} \quad K_k^b = \frac{p_{k|k-1}^b}{p_{k|k-1}^b + \sigma_{vb}^2(k)} \tag{11}$$

$$\text{Innovation:} \quad r_k^b = g_k^b - \hat{f}_{k|k-1}^b \tag{12}$$

$$\begin{aligned} \text{State update:} \quad \hat{f}_k^b &= \hat{f}_{k|k-1}^b + K_k^b r_k^b \\ &= K_k^b g_k^b + (1 - K_k^b)\hat{f}_{k-1}^b \end{aligned} \tag{13}$$

$$\text{Covariance update:} \quad p_k^b = (1 - K_k^b) p_{k|k-1}^b. \tag{14}$$

By examining the similarity between expressions (4) and (13), it is observed that the Kalman gain term K_k in (13) plays the same role as the linear filter coefficient ξ_k in (4). It is worth noting that the similarity between expressions (4) and (13) is a direct result of the system models defined in (7) and (8). Therefore, in Kalman filter based learning methods, the linear filter coefficient defined in (4) is given by

$$\boldsymbol{\xi}_k \equiv \mathbf{K}_k. \tag{15}$$

In essence, the Kalman filter adjusts its trust between the reference (f^b_{k-1}) and observed (g^b_k) data based on the estimated system noises (i.e., σ^2_{wb} and σ^2_{vb}). For example, small σ^2_{vb} tends to make K^b_k closer to one, thereby implying a higher trust in the observations g^b_k. In other words, the Kalman filter considers both internal variability (process noise) and external variability (observation noise) for appearance learning. Further, the Kalman gain is optimal in the sense of mean square error, under the previously stated Gaussian/linear assumptions.

However, the computation of the Kalman gains in (9–14) requires the knowledge of σ^2_{wb} and σ^2_{vb} which are usually unknown. This leads to the adaptive Kalman filter (AKF) that tries to estimate these noise variances. A brief overview of the different AKF methods was reported in [25]. A more recent survey was presented in [20, 31]. These methods are broadly divided in four categories, i.e., Bayesian, maximum likelihood, correlation and covariance methods. The former two have been deemed computationally expensive. In the following, two AKF-based appearance learning algorithms are presented that rely on covariance matching and correlation methods.

4.3 AKF Covariance Matching ($AKF_{\mathbf{cov}}$)

Covariance matching methods [25, 26] are based on the relation connecting the process, measurement noise variances and the covariance of the innovations. Since the innovations are observable as defined in (12), their covariance can be estimated empirically under suitable ergodic assumptions. Thus, if one of the two variances $\sigma^2_{wb}(k-1)$ and $\sigma^2_{vb}(k)$ is known, then the other can be estimated by matching the empirically estimated covariance to its theoretical value. Here, we adopt the specific technique used in [30, 34, 35] where $\sigma^2_{vb}(k)$ is assumed to be *known* and $\sigma^2_{wb}(k-1)$ is obtained by covariance matching.

The theoretical covariance of the innovation process for the system defined in (7) and (8) can be easily be derived as [25]

$$E[r^b_k r^b_j] = [p^b_{k-1} + \sigma^2_{vb}(k) + \sigma^2_{wb}(k-1)]\delta(k-j), \tag{16}$$

where $\delta(\cdot)$ is the Kronecker delta. With $\sigma^2_{vb}(k)$ known and p^b_{k-1} given by (14), an obvious empirical approach for solving $\sigma^2_{wb}(k-1)$ from (16) is to estimate $E[(r^b_k)^2]$ by computing the sample variance of (12) over the last L frames $\mathbf{y}_{k-L+1}, \ldots, \mathbf{y}_k$. However, because the process noise is time varying there is a delicate tradeoff between choosing L large enough to obtain statistically significant estimates while simultaneously choosing L small enough to track nonstationary changes in $\sigma^2_{wb}(k-1)$.

In existing literature, this issue has been addressed by assuming identical statistics across all bins of the histogram. This helps in increasing the sample size to larger than L while at the same time sampling from only the L most recent frames. In [35], it is assumed that $\sigma_{vb}^2(k)$ is independent of both b and k and that $\sigma_{wb}^2(k)$ is independent of b, so that all N_b bins of the histogram in each frame share identical noise variances. Similar assumptions on $\sigma_{vb}^2(k)$ are made for the template-based appearance model of [30], where b indexes pixels of a template rather than bins of a histogram. By assuming a common value $\sigma_{wb}^2(k-1)$ for all template pixels in the current frame, the innovations sample variance can be averaged across both pixels and time. In [34], the authors modeled $\sigma_{vb}^2(k)$, the observation noise, to be time varying and attributed it to the precision of template transformations in order to obtain an expression to explicitly evaluate it. Similar covariance matching was used to estimate the scale matrix in [29]. It is important to note that in addition to being able to estimate only one of the two unknown noise variances, the AKF$_{\text{cov}}$ method does not guarantee the estimated variance to be positive semi-definite (PSD).

In our setup, we assume that $\sigma_{vb}^2(k)$ is independent of both k and b and that $\sigma_{wb}^2(k)$ is independent of k and estimate $E[(r_k^b)^2]$ with the sample variance

$$\hat{C}_r(k) = \frac{1}{LN_b} \sum_{l=k-L+1}^{k} \sum_{b=1}^{N_b} (r_l^b)^2. \tag{17}$$

Under these assumptions, p_{k-1}^b is independent of b. Thus, we set $p_{k-1}^b = p_{k-1}^1, \forall b$; and use (17) in (16) to obtain the approximate solution

$$\sigma_{wb}^2(k-1) \approx \hat{C}_r(k) - \sigma_{vb}^2(k) - p_{k-1}^1. \tag{18}$$

As in [30, 35], the initialization at $k = 1$ is given by

$$\sigma_{vb}^2(k) = \frac{1}{2}\hat{C}_r(1)\,\forall b, k; \quad p_0^b = \frac{1}{2}\hat{C}_r(1)\,\forall b, \tag{19}$$

which implies $\sigma_{wb}^2(0) = 0$. We refer to this algorithm as AKF$_{\text{cov}}$ and use it in the following primarily as a baseline for comparison with the AKF$_{\text{als}}$ technique given in the next section.

4.4 AKF Autocovariance Based Least Squares (AKF$_{\text{als}}$)

When the filtering gain is not optimal, the residues will exhibit non-zero correlations at different lags. Autocovariance-based methods typically involve a set of constraints that relate the residual autocorrelations at different time lags with the unknown noise variances. Pioneering work in this field was reported by Mehra in [24, 25], where the residual autocorrelation is used for adaptive Kalman filtering. Mehra's method involves a three-step iterative process where a Lyapunov-type equation has to be solved at every iteration. To avoid solving the Lyapunov-type equation, Carew and Belanger [3] proposed a simpler procedure involving only matrix operations. Neethling and Young [28] pointed out that the methods in [3, 24, 25] yield estimates with large variances and do not consider the positive semi-definite (PSD) requirement for the unknown noise variances. Recently, Odelson et al. [33] presented an Autocovariance Least Squares (ALS) method which estimates both the unknown noise variances simultaneously and ensures that they are non-negative. In addition, estimates from the ALS method are more stable compared with Mehra's method and converge asymptotically to the optimal value with increasing sample size.

We next present the ALS method (AKF_{als}) in the context of histogram-based appearance learning. Consider the state space model given by (7) and (8). In the ALS approach, it is assumed that each bin in the histogram has unique noise characteristics. This implies that $\sigma^2_{wb}(k)$ and $\sigma^2_{vb}(k)$ depend on both k (non-stationary), b and that w_k^b and v_k^b are mutually uncorrelated. Thus, the noise statistics are different for each bin of the histogram and there is a separate coefficient ξ_k^b for each $b \in [1, N_b]$ differentiating it from the previously discussed methods and at the same time allowing for increased flexibility when adapting to appearance variations. In this work, the dependence on k is slightly relaxed to the case where the noise variances are assumed to be piecewise stationary over a small interval. Given an asymptotically converged Kalman gain in this interval K^b, the state estimate in (13) is given by

$$\widehat{f}_k^b = \widehat{f}_{k|k-1}^b + K^b r_k^b, \tag{20}$$

where the state estimation error $\varepsilon_k^b = f_k^b - \widehat{f}_k^b$ and the innovation (12) may then be formulated together into a state space model according to [32, 33]

$$\varepsilon_{k+1}^b = \overbrace{\left(1 - K^b\right)}^{\overline{A}^b}\varepsilon_k^b + \overbrace{\left[1 \ \ -K^b\right]}^{\overline{G}^b}\overbrace{\begin{bmatrix} w_k^b \\ v_k^b \end{bmatrix}}^{\overline{W}_k^b}, \tag{21}$$

$$r_k^b = \varepsilon_k^b + v_k^b. \tag{22}$$

Using the recursive nature of (21) and (22) to express the residue at time k, in terms of ε_0^b results in

$$r_k^b = (\overline{A}^b)^k \varepsilon_0^b + \sum_{m=0}^{k-1} (\overline{A}^b)^{k-m-1} \overline{G}^b \overline{W}_m^b + v_k^b. \tag{23}$$

Let

$$\mathscr{C}_j^b = E[r_k^b r_{k+j}^b] \tag{24}$$

be the innovations autocorrelation at lag j. We assume $E[\varepsilon_0^b] = 0$ and $\text{cov}(\varepsilon_0^b) = \pi_0^b$ and define

$$\overline{Q}_W^b \triangleq E\left[\overline{W}_k^b \overline{W}_k^{b\,T}\right] = \begin{bmatrix} \sigma_{wb}^2(k) & 0 \\ 0 & \sigma_{vb}^2(k) \end{bmatrix}, \tag{25}$$

$$\chi^b \triangleq E[\overline{W}_k^b v_k^b] = \begin{bmatrix} 0 \\ \sigma_{vb}^2(k) \end{bmatrix}. \tag{26}$$

Then we can obtain relations for the auto- and cross-correlation terms as:

$$\begin{aligned}
E[(r_0^b)^2] &= \pi_0^b + \sigma_{vb}^2(k),\\
E[r_1^b r_0] &= \overline{A}^b \pi_0^b + \overline{G}^b \chi^b,\\
E[r_2^b r_0] &= (\overline{A}^b)^2 \pi_0^b + \overline{A}^b \overline{G}^b \chi^b,\\
E[(r_1^b)^2] &= \overline{A}^b \pi_0^b \overline{A}^b + \overline{G}^b \overline{Q}_W^b \overline{G}^b + \sigma_{vb}^2(k),\\
E[r_2^b r_1] &= (\overline{A}^b)^2 \pi_0^b (\overline{A}^b)^2 + \overline{A}^b \overline{G}^b \overline{Q}_W^b \overline{G}^b + \overline{G}^b \chi^b,\\
E[(r_2^b)^2] &= (\overline{A}^b)^2 \pi_0^b (\overline{A}^b)^2 + \overline{A}^b \overline{G}^b \overline{Q}_W^b \overline{G}^b \overline{A}^b + \overline{G}^b \overline{Q}_W^b \overline{G}^b + \sigma_{vb}^2(k).
\end{aligned} \tag{27}$$

The heart of the ALS method lies in the fact that the residues r_k^b are statistically white in nature and are uncorrelated at different time lags if the Kalman gain is optimal. Therefore, ideally $E[r_k^b r_j^b]$ would be zero for all $k \neq j$. However, in practical situations this is not the case when the noise variances are unknown. The ALS method tries to estimate the unknown noise variances by exploiting their relationship with the non-zero cross-correlation values as in (27) where the left-side terms of the equation are estimated empirically from the observed residues. To formulate the problem in a least squares form an autocovariance matrix is defined $R^b(N_d)$. Here the matrix is shown for the case when $N_d = 3$ and can easily be generalized to arbitrary N_d

$$R^b(3) = E\begin{bmatrix} \left(r_k^b\right)^2 & r_k^b r_{k+1}^b & r_k^b r_{k+2}^b \\ r_k^b r_{k+1}^b & \left(r_{k+1}^b\right)^2 & r_{k+1}^b r_{k+2}^b \\ r_k^b r_{k+2}^b & r_{k+1}^b r_{k+2}^b & \left(r_{k+2}^b\right)^2 \end{bmatrix}. \tag{28}$$

Let I_3 be the 3×3 identity matrix, $\otimes$ denote the Kronecker product, and $\oplus$ denote direct sum. Let

$$\Theta^b = \begin{bmatrix} 1 \\ \overline{A}^b \\ (\overline{A}^b)^2 \end{bmatrix}, \quad \Gamma^b = \begin{bmatrix} 0 & 0 & 0 \\ 1 & 0 & 0 \\ \overline{A}^b & 1 & 0 \end{bmatrix}, \quad \Psi^b = \Gamma^b \left(\bigoplus_{i=1}^{3} K^b \right), \tag{29}$$

and let "vec" be the vectorization operator which transforms a matrix into a vector by stacking the columns upon one another. Then the vectorization of $R^b(3)$ is given by

$$\begin{aligned}\mathrm{vec}[R^b(3)] =& (\Theta^b \otimes \Theta^b)\pi_0^b \\ &+ \Gamma^b \otimes \Gamma^b \,\mathrm{vec}(I_3)\,\mathrm{vec}\left(\overline{G}^b \overline{Q}_W^b \overline{G}^{b^T}\right) \\ &+ (\Psi^b \oplus \Psi^b + I_3)\,\mathrm{vec}(I_3)\sigma_{vb}^2(k).\end{aligned} \tag{30}$$

Let

$$D^b = (\Theta^b \otimes \Theta^b)\left(1 - \overline{A}^b \otimes \overline{A}^b\right)^{-1} + (\Gamma^b \otimes \Gamma^b)\,\mathrm{vec}(I_3). \tag{31}$$

Upon eliminating π_0^b from (30) and substituting for $\overline{Q}_w^b$, we obtain

$$\overbrace{\mathrm{vec}[R^b(3)]}^{\mathscr{C}^b} = \overbrace{\left[\, D^b |\, D^b K^{b^2} + (\Psi^b \oplus \Psi^b + I_9)\,\mathrm{vec}(I_3) \right]}^{\mathscr{A}^b} \overbrace{\begin{bmatrix} \sigma_{wb}^2(k) \\ \sigma_{vb}^2(k) \end{bmatrix}}^{\mathbf{x}^b}. \tag{32}$$

The expression (32) forms the core of the ALS method as it relates the correlation terms $\mathscr{C}_j$ to the unknown noise covariances σ_w^2 and σ_v^2. Under the reasonable assumption that the innovations process is locally ergodic, the quantity $\mathscr{C}^b$ in (32) can be approximated using the residues computed from the filter according to

$$\hat{\mathscr{C}}_j^b = \frac{1}{N_d - j} \sum_{i=1}^{N_d - j} r_i^b r_{i+j}^b. \tag{33}$$

We have (32) in the form $\mathscr{A}^b \mathbf{x}^b = \widehat{\mathscr{C}^b}$. Thus, the least squares problem for the unknown noise variances $\sigma_{wb}^2(k)$ and $\sigma_{vb}^2(k)$ can be expressed as

$$\Phi^b = \min_{\sigma_{wb}^2(k), \sigma_{vb}^2(k)} \left\| \mathscr{A}^b \begin{bmatrix} \sigma_{wb}^2(k) \\ \sigma_{vb}^2(k) \end{bmatrix} - \widehat{\mathscr{C}^b} \right\|^2 \tag{34}$$

subject to $\sigma_{wb}^2(k), \sigma_{vb}^2(k) \geq 0$. The positive semidefinite requirements on $\sigma_{wb}^2(k)$ and $\sigma_{vb}^2(k)$ are enforced by appending a logarithmic barrier function to (34), resulting in

$$\begin{aligned}\Phi^b =& \min_{\sigma_{wb}^2(k), \sigma_{vb}^2(k)} \left\| \mathscr{A}^b \begin{bmatrix} \sigma_{wb}^2(k) \\ \sigma_{vb}^2(k) \end{bmatrix} - \widehat{\mathscr{C}^b} \right\|^2 \\ &- \mu \log[\sigma_{wb}^2(k)\sigma_{vb}^2(k)],\end{aligned} \tag{35}$$

where μ is the barrier parameter. The least squares problem (35) has been shown to be convex and can be solved using a Newton recursion [33]. Pseudo-code to implement the $\mathrm{AKF}_{\mathrm{als}}$ algorithm for a single bin of the histogram is given in Table 1.

Table 1 Pseudo-code of AKF_{als} for a single bin of the histogram at time k

1. Predict bin value $\widehat{f}^b_{k|k-1} = \widehat{f}^b_{k-1}$.
2. Acquire observation g^b_k based on tracker output.
3. Compute innovation $r^b_k = g^b_k - \widehat{f}^b_{k|k-1}$.
4. for $j = 0$ to $N_d - 1$
   Compute $\widehat{\mathscr{C}}^b_j = \frac{1}{L-j}\sum_{i=k-L+1}^{k-j} r^b_i r^b_{i+j}$.
   end
5. Optimize ALS problem to obtain $\sigma^2_{vb}(k)$ and $\sigma^2_{wb}(k)$.
6. Compute Kalman gain K^b_k using estimated noise variances.
7. Update bin value $\widehat{f}^b_k = \widehat{f}^b_{k|k-1} + K^b_k r^b_k$.

4.5 Comparison Between AKF_{als} and AKF_{cov}

The AKF_{cov} method makes two important but potentially problematic assumptions: (1) all histogram bins share same noise statistics, and (2) the variance of measurement noise is a known constant. In addition, there is no guarantee to ensure that the estimation of (16) always result in positive values for σ^2_w, the process noise variance. Since p_{k-1} computed in (16) is only an approximation of the actual error covariance, convergence of σ^2_w to the optimal value cannot be guaranteed. Unlike the AKF_{cov} method, the AKF_{als} method (1) estimates both process and observation noise parameters simultaneously, (2) computes noise statistics for each individual bin of the histogram, (3) enforces the estimated noise variances to be positive, and (4) is based on multiple constraints by considering the autocorrelation of the residues at different time lags. Our simulations show that the AKF_{als} method significantly outperforms the AKF_{cov} method in terms of the accuracy of noise estimation for linear systems with stationary noise processes.

The ALS method in [33] was originally developed for an offline application, where sufficient measurement data are available beforehand. Further, both the AKF_{cov} and AKF_{als} methods are derived under the assumption that the noise processes are stationary. However, in the case of infrared tracking, it is imperative that the histogram-based appearance be learnt "on-the-fly" and the noise processes affecting each histogram bin are not necessarily stationary. In this work, the noise processes are assumed to be piecewise stationary within a windowed time period. We apply the ALS method to residues obtained over a window of fixed width (N), implying that the target appearance is updated every N frames based on the observed variation.

To examine the validity of the above assumptions, we performed some numerical experiments with a simulated linear system where the noise processes were set to be non-stationary by allowing both step and continuous variations in terms of their variances. It was found that the AKF_{als} method was able to cope with the non-stationarity, and the actual performance depends on the chosen window size. On the one hand, a smaller window size allows for rapid adaptation to the noise variation, but tends to have a more unstable noise estimation. On the other hand, a larger window size improves noise estimation, but it is slow to adapt to noise variations. Therefore, the window size strikes a crucial balance between the reliability and sensitivity of appearance learning and must be chosen to maintain a good balance between the two. Furthermore, the estimates obtained by the AKF_{cov} method were found to have significantly larger errors compared to the AKF_{als} method. Our experiments show that a system that has slow-varying noise statistics can be approximated as a piecewise stationary system where the AKF_{als} method can be used to estimate the system and measurement noise variances in a piecewise manner.

5 Particle Filter-Based Tracking

This section details a particle filtering-based tracking algorithm where histogram based appearance learning is involved. We first present the target appearance model used for FLIR tracking, followed by the dynamics used for target tracking. The implementation of the algorithm is also discussed.

5.1 Dual Foreground–Background Appearance Model

In most FLIR sequences, the initial target is usually small (3–10 pixels) and its appearance is often blended with background. Motivated by the "hit-or-miss" morphological transform that uses both foreground and background for object detection, we present a target model that involves the local statistics of both the target and its surrounding areas. Let $\mathbf{y}_k$ represent the k th frame, and $\mathbf{x}_k = [x_k, y_k; s_k^x, s_k^y]$ the state to be estimated, where (x_k, y_k) and (s_k^x, s_k^y) are the position (top-left corner) and size, respectively. As shown in Fig. 4, the target appearance, denoted by $\mathbf{G}(\mathbf{x}_k)$, is composed of four histograms: the foreground/background intensity $f_{fi}(\mathbf{x}_k)/f_{bi}(\mathbf{x}_k)$, foreground/background stdev $f_{fs}(\mathbf{x}_k)/f_{bs}(\mathbf{x}_k)$, which are extracted from $\mathbf{y}_k$ by using the kernel-based method in [5, 43]:

$$\mathbf{G}(\mathbf{x}_k) = \{f_{fi}(\mathbf{x}_k), f_{bi}(\mathbf{x}_k), f_{fs}(\mathbf{x}_k), f_{bs}(\mathbf{x}_k)\}. \tag{36}$$

Given a hypothetical target area $\mathbf{x}_k$ in frame $\mathbf{y}_k$, we can obtain its corresponding appearance hypothesis as $\mathbf{G}(\mathbf{x}_k)$, and then the similarity between $\mathbf{G}(\mathbf{x}_k)$ and the reference model $\mathbf{F}_{k-1}$ is computed as

$$\text{D}(\mathbf{G}(\mathbf{x}_k), \mathbf{F}_{k-1}) = \sum_{z \in Z} v_z \cdot \text{d}(f_z(\mathbf{x}_k), f_{z,k-1}), \tag{37}$$

where $Z = \{fi, bi, fs, bs\}$ and d is defined in (5). The v_z's are also used to adjust the relative significance of the four histograms and may be adaptively selected as

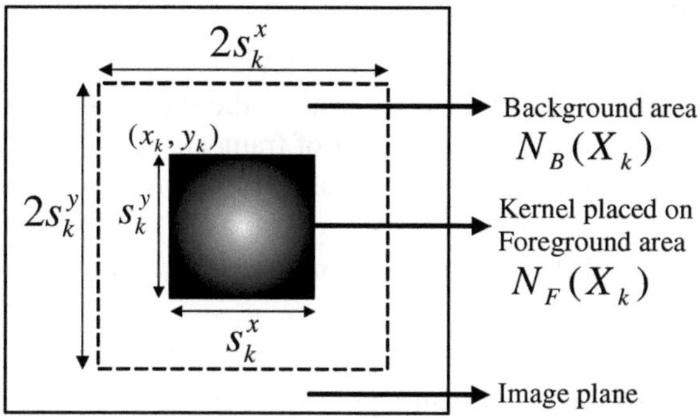

Fig. 4 Foreground region $N_F(\mathbf{x}_k)$ with overlapped kernel and background area $N_B(\mathbf{x}_k)$ defined based on $\mathbf{x}_k = [x_k, y_k, s_k^x, s_k^y]$

discussed in [43]. In this work, all four histograms are given equal importance during tracking for the AMCOM sequences, i.e., $v_{fi} = v_{bi} = v_{fs} = v_{bs} = 0.25$.

5.2 Target Dynamics

Two dynamic models are needed for FLIR tracking, one each for the position and size. In most FLIR sequences, the target predominantly exhibits relatively stable ground motion accompanied by strong ego-motion of the airborne sensor. Previous works in [8, 49] used a separate global motion model to compensate the sensor ego-motion. Inspired by [21], we use an adaptive motion model to capture both ground motion and ego-motion for FLIR tracking:

$$\mathrm{x}_k = \mathrm{x}_{k-1} + C_k v_k, \tag{38}$$

where $\mathrm{x}_k = [x_k, y_k]$, $C_k \propto \mathbf{E_n}[\Delta \mathrm{x}_k]$ and $v_k \sim N(0, I)$. $\mathbf{E_n}[\Delta \mathrm{x}_k]$ is the estimated velocity (in the image plane) estimated over the past n frames. In essence, this model controls the spread of particles in proportion with the observed target velocity.

Due to the nature of range closure sequences, the target size usually increases slowly when the target is small and is followed by rapid amplification as the sensor moves closer to the target. To account for such size changes, a simple model that can increase or decrease the size by up to 20% at each time step is advocated. The state transition model for the size vector s_k is defined as

$$\mathrm{s}_k = D\mathrm{s}_{k-1}, \tag{39}$$

where $\mathrm{s}_k = [s_k^x, s_k^y]$, $D \sim U(0.8, 1.2)$. This model controls size change in proportion with the previous size.

Table 2 Pseudo-code of the particle filter algorithm with online appearance learning

- Initialization: Draw $\mathbf{x}_1^j \sim N(X_1, 1), \forall j = \{1, \cdots, N_p\}$ and set $\mathbf{F}_1 = \mathbf{G}(X_1)$, where X_1 is the ground truth of the state in the initial frame.
- For $k = 2, \cdots, T$ (number of frames)
  1. For $j = 1, \cdots, N_p$ (number of particles)
     1.1 $\mathbf{x}_k^j \sim p(\mathbf{x}_k^j|\mathbf{x}_{k-1}^j)$ using (39) and (38)
     1.2 Compute $w_k^j = p(\mathbf{y}_k|\mathbf{x}_k^j, \mathbf{F}_{k-1})$ using (40)

     End
  2. Normalize the weights such that $\sum_{j=1}^{N_p} w_k^j = 1$.
  3. Compute the mean of the states $\hat{\mathbf{x}}_k = \sum_{j=1}^{N_p} w_k^j \mathbf{x}_k^j$.
  4. Set $\mathbf{x}_k^j$ =resample$(\mathbf{x}_k^j, w_k^j)$.
  5. Update reference model $\mathbf{F}_k$ based on state estimate $\hat{\mathbf{x}}_k$.
- End

5.3 Target Tracking with Appearance Learning

We develop a SIR (sequential importance re-sampling)-based tracking algorithm that involves three steps: *particle propagation*, *particle evaluation*, and *appearance learning*. The complete tracking algorithm is shown in Table 2. First, particles are drawn according to the state dynamics defined in (38) and (39). Second, particle weights are computed by the likelihood function $p(\mathbf{y}_k|\mathbf{x}_k, \mathbf{F}_{k-1})$, which is defined based on the distance measure in (37) as

$$p(\mathbf{y}_k|\mathbf{x}_k, \mathbf{F}_{k-1}) \propto \exp(\lambda \cdot \mathrm{D}(\mathbf{G}(\mathbf{x}_k), \mathbf{F}_{k-1})), \tag{40}$$

where λ is a constant, $\mathbf{G}(\mathbf{x}_k)$ is the hypothesized target appearance extracted from the kth frame $\mathbf{y}_k$, and $\mathbf{F}_{k-1}$ is the previous reference model. After normalization and re-sampling, we can obtain new state estimates based on which the appearance learning is invoked according to Table 1.

6 Detecting Occlusions and Track Losses

The task of excluding outlier observations is quite common in Kalman filtering applications and is referred to as *gating*. The gating criterion is often based on the statistical properties of the residues. In [29] an error norm was defined on the residues to prevent outlier pixels from corrupting the appearance information. In [35], a hypothesis test was defined based on the standard deviation of the residues. This was possible because residues from each bin of the histogram were assumed to be statistically similar. Other rigorous methods that employ more sophisticated

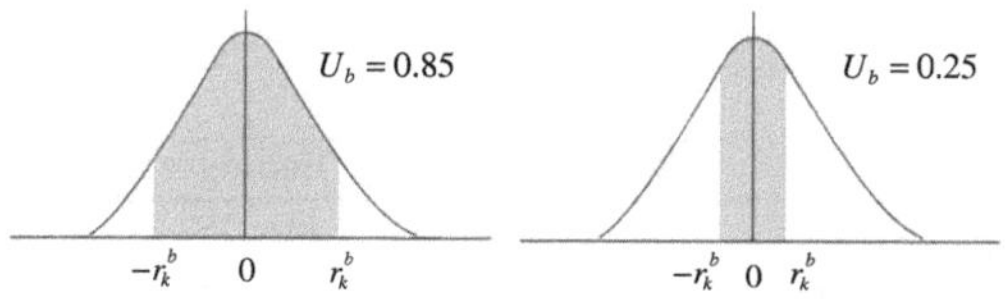

Fig. 5 Illustration of the uncertainty associated with two different residual values for the underlying distribution $N(0, \sigma_{vb}^2)$

statistical tests are discussed in [38]. In this work, we make use of the fact that the residues computed in (12) for each histogram bin b can be approximated by a Gaussian distribution $r_k^b \sim N(0, \sigma_{vb}^2)$ if the prediction error covariance is negligible. Note that the residues indicate the degree of the mismatch between the predicted and observed bin value. Since each of the histogram bin in AKF_{als} is characterized separately, we associate each bin with an uncertainty term U_b defined as

$$U_b = \frac{1}{\sqrt{2\pi\sigma_{vb}^2(k)}} \int_{-r_k^b}^{r_k^b} \exp\left(\frac{-x^2}{2\sigma_{vb}^2(k)}\right) dx. \tag{41}$$

Ideally if the observed appearance exactly matches the reference model ($r_k^b = 0$), then there is very little chance of the observation being erroneous. The uncertainty values for two different r_k^b's for the same underlying distribution are shown in Fig. 5. We see that a value of r_k^b closer to zero results in lower uncertainty and a value farther away leads to higher uncertainty. During the filter operation we refrain from updating the appearance model when the average uncertainty over the non-zero bins rises above a threshold of 0.7 and declare a temporary "track loss". The value of 0.7 was selected based on experiments on the VIVID dataset. A prolonged track loss state is indicative of an occlusion or movement of the target outside the sensor view.

Once in lost mode, in each subsequent frame, attempts are made to reacquire the target by using a histogram matching based detector. The detection is performed in a window area around the last known target position and an average of the last few appearance histograms is used as the reference. Being dependent on only the last known appearance can be problematic in cases of partial occlusion because a significant portion of the target appearance may already be lost/occluded before a track loss is detected. The target is declared as found and normal tracking resumes when the detected candidate region has a uncertainty value below the threshold 0.7. If no acceptable candidates are found within 100 frames of a temporary track loss, a complete "track loss" is declared and tracking is terminated. This simple methodology works very well in recovering the target after short periods of occlusions and scene absence and is discussed further in the experiments section. In case of a prolonged absence from view or occlusions, when the target reappears it may significantly differ from its previous known appearance. This makes it difficult for the detector to identify the target with high confidence and may require re-initialization.

Table 3 List of the AMCOM sequences used in experiments

Sequences	Frame			Size	
	Starting frame	Ending frame	Length	Starting size	Ending size
LW-15-NS	20	270	250	5×8	16×16
LW-17-01	1	350	350	5×8	16×29
LW-21-15	235	635	400	3×4	10×10
LW-14-15	1	225	225	4×5	23×19
LW-22-08	50	300	250	5×8	17×24
LW-20-18	120	420	300	4×7	10×17
LW-18-17	1	190	190	5×9	11×25
LW-19-06	40	260	220	3×4	6×11
MW-14-10	1	450	450	6×11	12×28
LW-20-04	10	360	350	3×4	12×15

7 Experimental Results

Our tracking algorithm was tested on the AMCOM FLIR dataset and the VIVID dataset. We use the AMCOM dataset to show the merits of the appearance learning of our algorithm since the sequences in the dataset exemplify the challenges of FLIR tracking such as poor target visibility, strong ego-motion, small targets, size variations, dust clouds, significant clutter and background noise. Ground truth information about the target position, size and type is available in the dataset and serves as a benchmark for performance evaluation. Ten representative FLIR sequences used in the experiment are given in Table 3. And the sequences in the VIVID dataset present frequent occlusion by foliage, target exiting and re-appearing, which are good to demonstrate the performance of track loss detection component of our algorithm.

7.1 Experiments on the AMCOM Dataset

7.1.1 Experimental Setup

Three appearance learning algorithms, namely HS, AKF_{cov} and AKF_{als} are integrated with the same tracking algorithm given in Table 2. It is worth mentioning that all three algorithms share the same linear filtering form defined in (4). HS determines ξ_k according to histograms similarity, while AKF_{cov} and AKF_{als} use the Kalman gain. The detailed setting of the three tracking algorithms is listed in Table 4. In practice, the appearance learning algorithms were applied only to the intensity histograms, as the dynamics of stdev histograms do not have a well-defined structure. The stdev histograms in all cases were updated using the HS method.

Table 4 Description and value of the experimental parameters

Variables	Description	Values
$N_b^{(1)}$	The bin number of the intensity histogram	32
$N_b^{(2)}$	The bin number of the stdev histogram	16
L	The number of frames used for AKF_{cov} in (17)	3
N	The number of frames used for obtaining residues in AKF_{als}	7
N_d	The number of autocorrelation lags considered in AKF_{als}	5
C_k	Dynamics of position used in (38)	$3\mathbf{E_n}[\triangle \mathrm{x}_k]$
N_p	The number of particles used for tracking	200

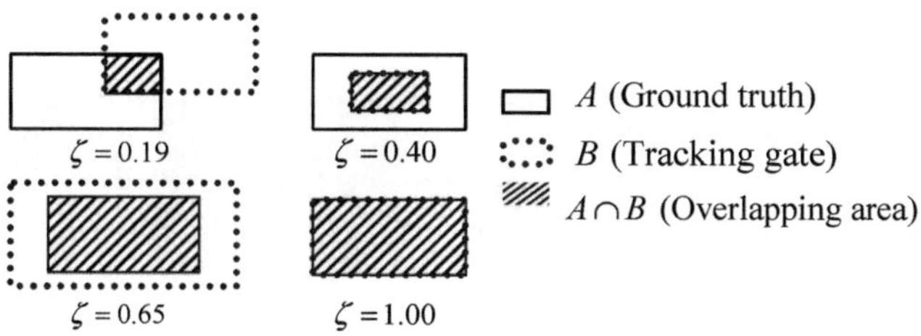

Fig. 6 Illustration of the overlap metric for a few different tracking cases

In addition to the tracking errors, we adopt an overlap metric proposed in [40] to quantify the overlap between the tracking gate with the actual target. Let A and B represent the tracking gate and the ground-truth bounding box respectively, then the overlap ratio ζ is defined as

$$\zeta = \frac{\#(A \cap B) \times 2}{\#(A) + \#(B)}, \tag{42}$$

where # is the number of pixels. A few examples about this metric are shown in Fig. 6.

7.1.2 Experimental Analysis

Three tracking algorithms (50 Monte Carlo runs) were evaluated and compared in terms of their appearance learning performance (Fig. 7), the overlap metric ζ (Fig. 8) and the tracking error (Fig. 10 and Table 6).

Appearance learning. As shown in Fig. 7, it can easily be observed that the results of AKF_{als} closely match the ground-truth. Closer examination reveals that HS and AKF_{cov} result in the histograms that slowly deviate or "drift" from the true ones. This is clearly evident in Fig. 7c, where the intensity variation in the latter part of the sequence (around frame 300) is not captured by HS and AKF_{cov}. Therefore, the tracker includes a large portion of the background into the tracking gate as seen in frames 320, 360 of Fig. 10(row 3).

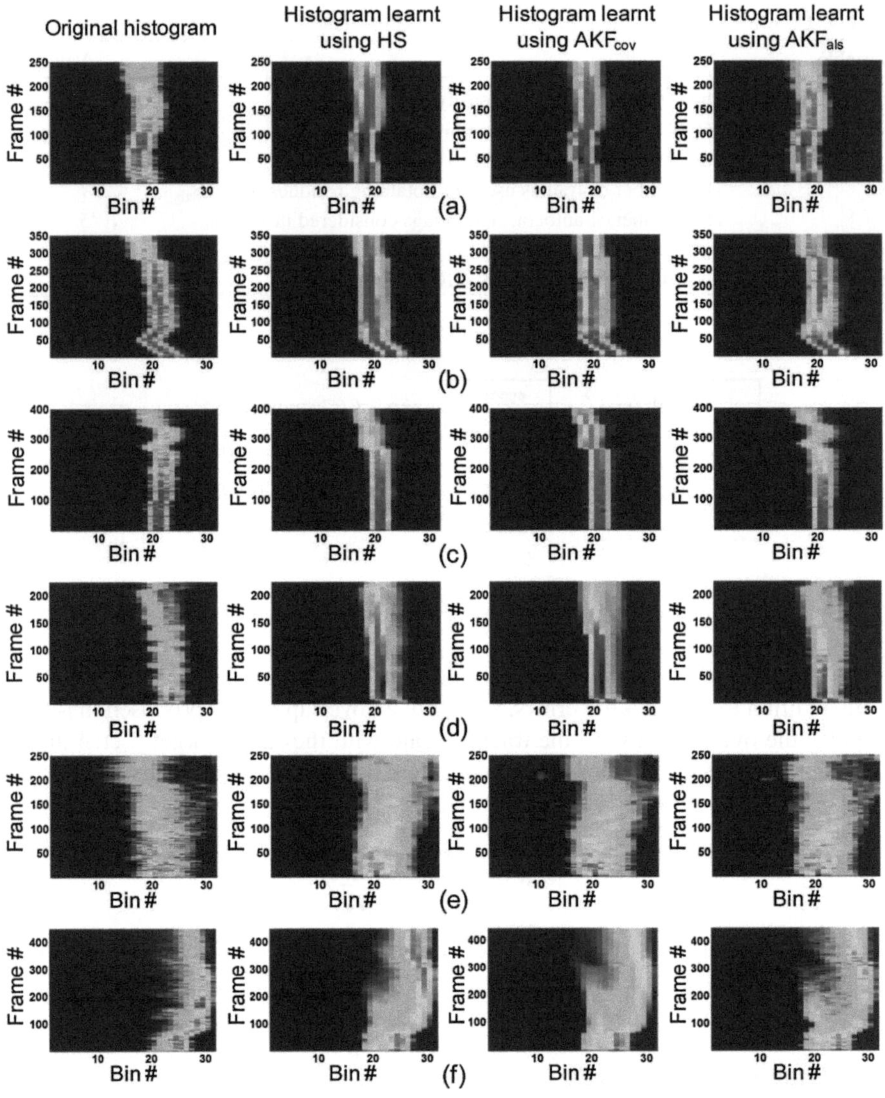

Fig. 7 Comparison of appearance learning for three sequences: **(a)** LW-15-NS; **(b)** LW-17-01; **(c)** LW-21-15; **(d)** LW-14-15; **(e)** LW-22-08 and **(f)** MW-14-10

Overlap metric. The improvements of appearance learning can be further reflected by the overlap metric in Fig. 8b, which compares ζ_{als}, ζ_{cov} and ζ_{HS} in pairwise. For example, the improvement of AKF$_{als}$ over AKF$_{cov}$ or HS can be demonstrated by seeing most data points are above the diagonal lines. The comparable result of AKF$_{als}$ and AKF$_{cov}$ in sequence LW-22-08 is also shown in the similar appearance learning performance in Fig. 7e where the histogram-based appearance lacks strong modes

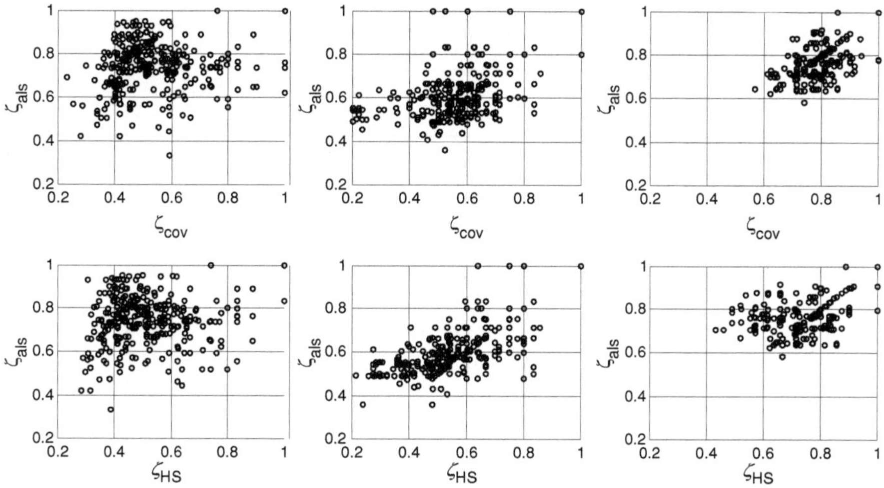

(a) LW-17-01 (b) LW-21-15 (c) LW-22-08

Fig. 8 Comparison of the overlap metric ζ_{als} with ζ_{cov} (*top*) and ζ_{HS} (*bottom*) plotted over the entire length of three sequences, LW-17-01, LW-21-15, LW-22-08

Table 5 The overlap metric values of the three tracking algorithms

Sequences	HS	AKF_{cov}	AKF_{als}
LW-15-NS	0.669	0.707	*0.714*
LW-17-01	0.547	0.596	*0.720*
LW-21-15	0.601	0.578	*0.620*
LW-14-15	0.676	0.682	*0.708*
LW-22-08	0.751	*0.770*	0.758
LW-20-18	0.689	0.753	*0.758*
LW-18-17	*0.704*	0.702	0.703
LW-19-06	0.670	0.685	*0.713*
MW-14-10	*0.802*	0.797	0.799
LW-20-04	0.715	0.711	*0.720*
Average	0.682	0.698	*0.721*

and has widespread and small bin values. The average value of ζ corresponding to different algorithms is given in Table 5. The AKF_{als} method has the largest value that indicates its superior performance of target tracking when compared to the other two algorithms.

Tracking error. Table 6 provides quantitative results of the tracking performance. In most cases, AKF_{als} produces the least errors in terms of both position and size. The HS approach loses track of the target in sequences LW-20-18 (6 runs) and LW-19-06 (2 runs) as indicated by the large errors. The AKF_{cov} also loses track of the target in the sequence LW-20-18 (1 run) due to the high similarity between

Table 6 The mean error of the state variables over averaged over the length of the sequence from 50 Monte Carlo runs using three different algorithms

Algorithm	HS				AKF_{cov}				AKF_{als}			
Sequences	x	y	s^x	s^y	x	y	s^x	s^y	x	y	s^x	s^y
LW-15-NS	1.02	1.82	1.91	2.73	0.86	1.51	1.64	2.40	*0.80*	*1.46*	*1.42*	*2.34*
LW-17-01	2.41	3.42	2.10	*3.02*	2.15	3.01	2.10	3.16	*1.21*	*2.11*	*1.38*	3.03
LW-21-15	0.97	1.65	*2.62*	2.94	1.14	1.81	2.80	3.11	*0.89*	*1.30*	2.79	*2.58*
LW-14-15	*0.89*	0.82	3.16	2.14	0.93	*0.79*	2.98	2.16	1.10	0.80	*2.66*	*1.79*
LW-22-08	*1.17*	0.87	1.68	*2.05*	1.20	0.84	*1.07*	2.23	1.20	*0.84*	1.36	2.18
LW-20-18	3.23	1.83	1.66	1.95	0.90	1.10	*1.31*	1.77	*0.60*	*1.08*	1.44	*1.75*
LW-18-17	*1.27*	1.72	*0.73*	2.95	1.30	1.84	0.86	2.61	1.43	*1.68*	1.09	*2.25*
LW-19-06	1.98	1.55	1.57	1.54	0.80	0.76	1.68	1.45	*0.69*	*0.71*	*1.54*	*1.28*
MW-14-10	*0.63*	0.79	1.65	1.69	0.76	0.81	1.64	1.79	0.78	*0.78*	*1.63*	*1.61*
LW-20-04	0.70	0.95	*0.94*	1.53	0.70	0.94	1.07	1.61	*0.69*	*0.91*	1.01	*1.36*
Average	1.43	1.54	1.80	2.25	1.07	1.34	1.72	2.23	*0.94*	*1.17*	*1.63*	*2.02*

foreground and background. More visual comparisons are shown in Fig. 10. We can see that AKF_{als} offers the best position and size estimation except sequence LW-22-08, where AKF_{cov} is slightly better due to the lack of well defined structure in the histogram-based appearance, as shown in Fig. 7e.

7.1.3 Tracking Performance of Covariance Descriptor

We also tested the covariance descriptor for FLIR tracking. The covariance descriptor was found to be robust and effective for object tracking in optical images and plays an important role in several state-of-the-art tracking algorithms [19, 37]. It was first proposed in [42] for object detection. This descriptor has several advantages: (1) it is able to fuse together many different features; (2) it is invariant to illumination conditions and rotation, contains both statistical and spatial information and is fast to compute; (3) it can be updated incrementally and systematically by some manifold learning methods. In infrared tracking, the covariance descriptor involves local intensity, stdev, gradient, orientation and Laplacian information of the target area. This descriptor was combined with a particle filter whose dynamics were described in Sect. 5.

The tracking results of using the covariance descriptor are shown in Fig. 9 where no learning is involved. It is observed that this covariance tracker is able to maintain a reasonable track of the target in LW-17-01, but fails to track the dark target in LW-15-NS. In both sequences, the tracker encounters difficulty in size estimation. The small size of the target, weak texture and absence of color significantly reduce the effectiveness of the covariance descriptor for tracking small targets in FLIR images. For the same reasons, the learning of the covariance descriptor using the method in [19] did not considerably improve the tracking results in FLIR imagery.

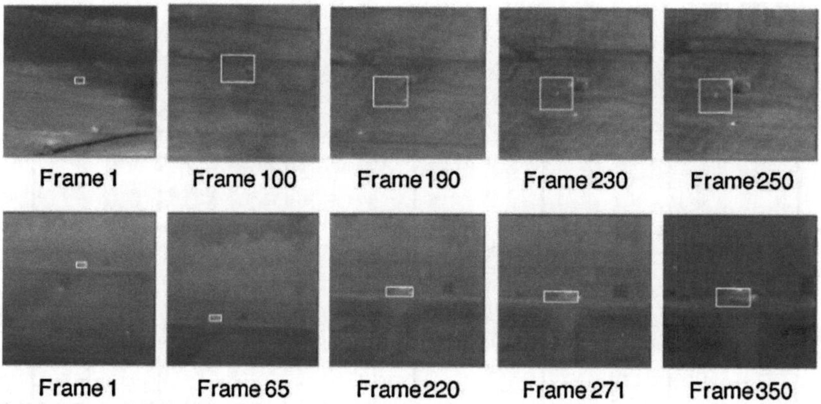

Fig. 9 Tracking results for two AMCOM sequences using the covariance tracker. *Top*: LW-15-NS and *bottom*: LW-17-01

7.1.4 More Discussion

In summary, the HS method is usually encumbered by the drifting problem during incremental appearance learning. The AKF_{cov} method, which assumes the same noise statistics for all bins in a histogram and estimates only the process noise without considering PSD conditions, results in a suboptimal Kalman gain. Its performance is marginally better than that of HS. The AKF_{als} algorithm, which estimates both process and observation noises with PSD constraints for each individual bin in a histogram, is able to follow the dynamics of the histogram during tracking and supports effective appearance learning. However, when a histogram is less structured or has many small bin values, such as LW-22-08 and MW-14-10, all three methods are comparable. This is mainly because the poor histogram structure may invalidate the Kalman filter assumptions. Then HS can incorporates the most recent tracker's observation for appearance learning when the histogram is less well defined. This justifies the use of HS for learning the stdev histograms which normally have weak structures.

7.2 Experiments on the VIVID Dataset

In the VIVID dataset the targets are larger compared to the AMCOM dataset and the foreground information is usually sufficient to represent a target. Therefore we predominantly depend on the foreground information for tracking by setting the histogram importance as $\upsilon_{fi} = 0.45$, $\upsilon_{bi} = 0.05$, $\upsilon_{fs} = 0.45$ and $\upsilon_{bs} = 0.05$ in (37). The sequences tested are typical of aerial surveillance videos and are affected by ego-motion of the sensor, occlusion by foliage, targets exiting scene and reappearing. Due to the absence of explicit ground truth information we only present visual

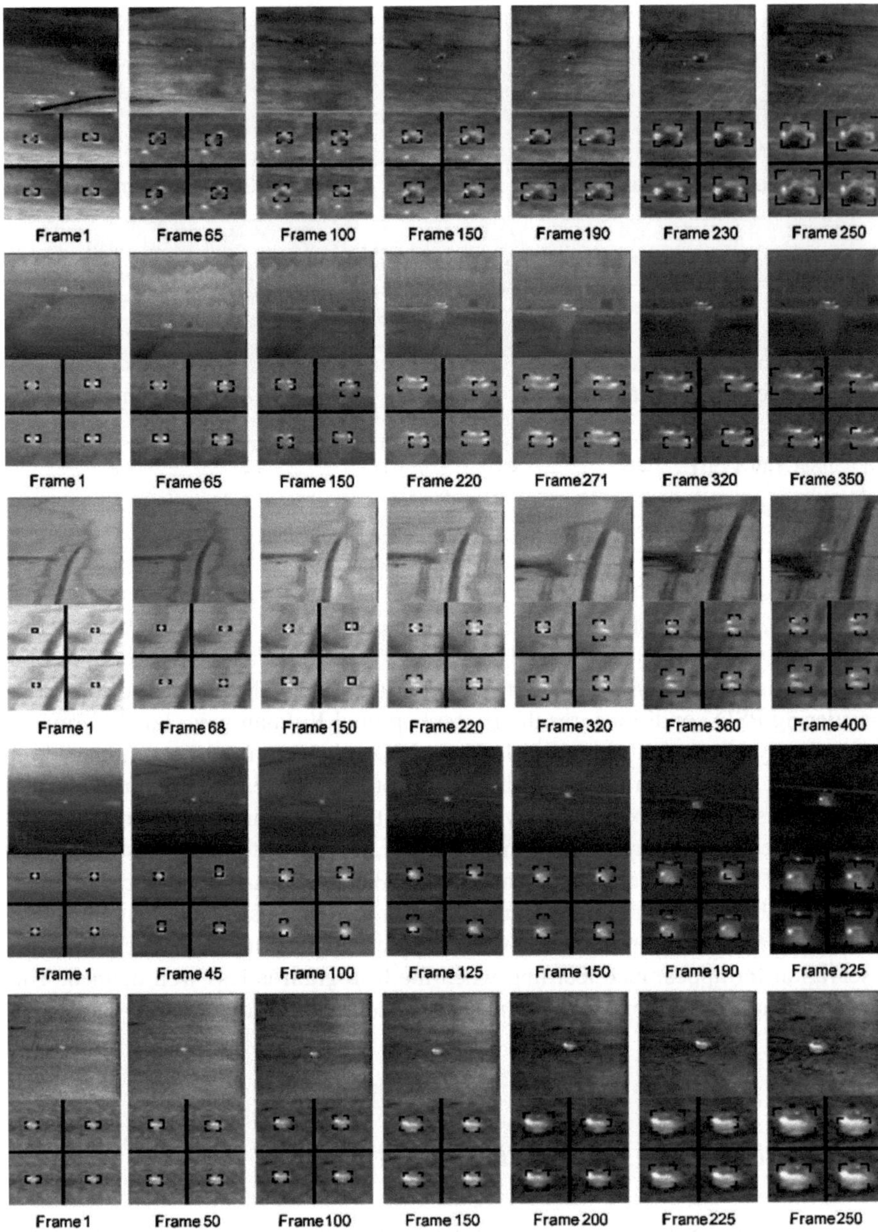

Fig. 10 Tracking results of the three algorithm on five AMCOM sequences. The *top row* of each image shows the observed frame and the *bottom row* depicts the tracking gates corresponding to the Ground truth (*top-left*), HS (*top-right*), AKF_{cov} (*bottom-left*), AKF_{als} (*bottom-right*). The sequences from *top* to *bottom* are LW-15-NS, LW-17-01, LW-21-15, LW-14-15 and LW-22-08

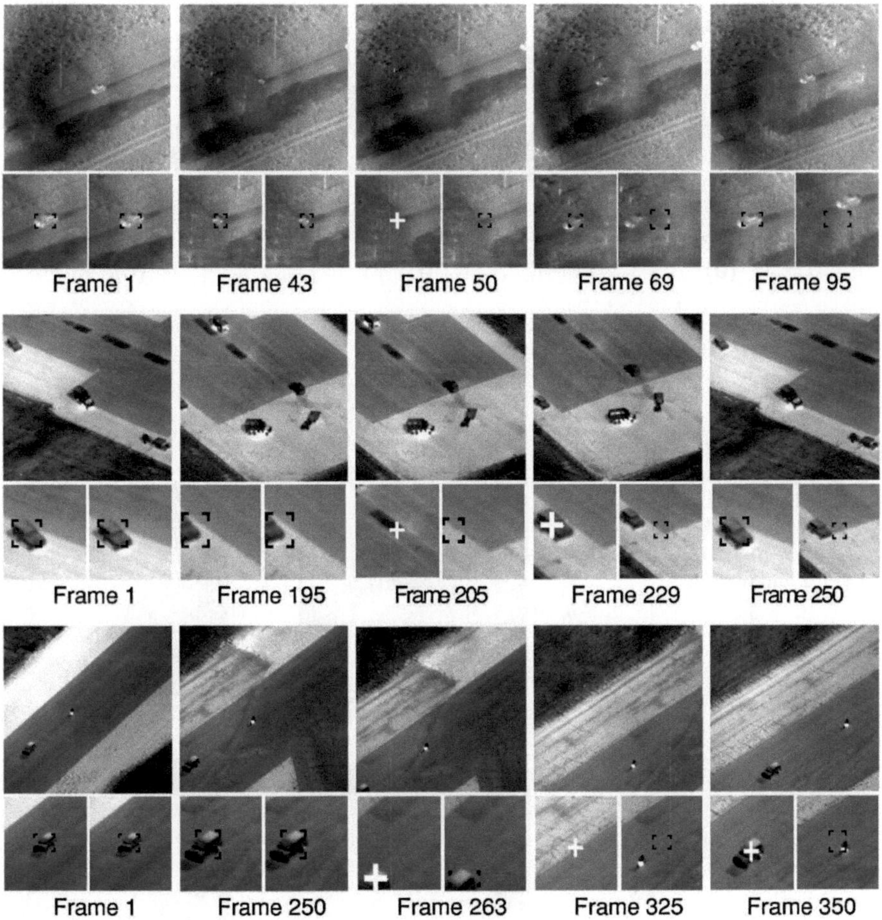

Fig. 11 A few sample frames from three different sequences are shown in each of the rows. In each sub-image, the *top* image represents the observed frame. *Bottom left* is the result of $AKF_{als+tld}$ and *bottom right*: AKF_{als}. The *black bounding boxes* represent the tracking result. The *white* "+" *sign* represents the output of the detector in the $AKF_{als+tld}$ method

evidence of the tracking performance. In all the sequences, the targets were manually initialized with an appropriate bounding box. We compare the performance of the tracking algorithm with ($AKF_{als+tld}$) and without (AKF_{als}) track-loss detection on a few representative sequences.

A few sample frames from three different sequences, the uncertainty associated with the $AKF_{als+tld}$ tracker and the corresponding foreground appearance variations are shown in Figs. 11, 12 and 13 respectively. In SEQ1 corresponding to the top row of Fig. 11, the target is occluded by some trees around frame 50 and reemerges around frame 65. Both the $AKF_{als+tld}$ and AKF_{als} algorithms perform similarly till the time of occlusion. During the period of occlusion the uncertainty associated

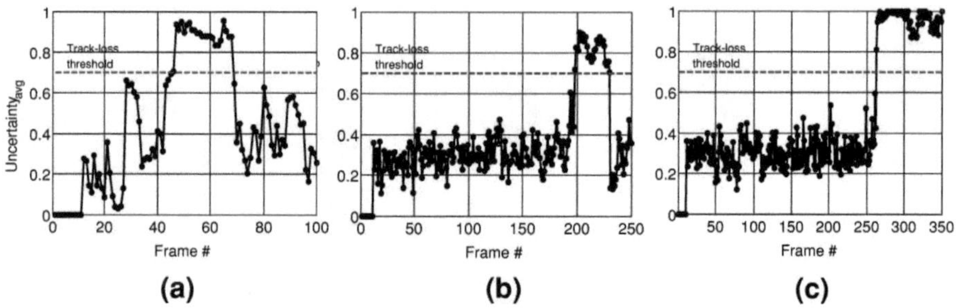

Fig. 12 The appearance uncertainty associated with the foreground histogram by the $AKF_{als+tld}$ tracker for **(a)** SEQ1, **(b)** SEQ2 and **(c)** SEQ3

Fig. 13 Foreground histogram variation associated with the $AKF_{als+tld}$ and AKF_{als} trackers for **(a)** SEQ1, **(b)** SEQ2 and **(c)** SEQ3

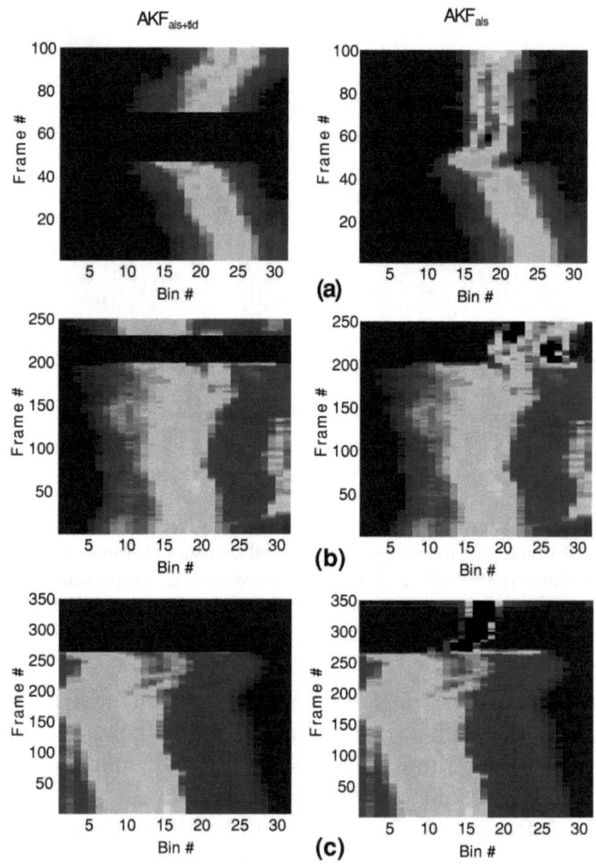

with the appearance increases above the set threshold as shown in Fig. 12a. Therefore, the $AKF_{als+tld}$ tracker stops all update to the appearance and goes into detection mode. The AKF_{als} tracker on the other hand continues to learn new appearance corresponding to the background. By the time the target is occlusion free, the detector is able to locate the target with acceptable level of certainty and the $AKF_{als+tld}$ tracker begins tracking the target. The AKF_{als} tracker by this time has learnt the appearance of the background and loses track of the target. From Fig. 13 we can observe the appearance of significant peaks in the appearance histogram of the AKF_{als} tracker as it shifts its focus toward the background. The $AKF_{als+tld}$ tracker suspends all updates until a reliable target signature is found, this is indicated by the missing values between the occlusion frames in Fig. 13a. In this sequence, the increased uncertainty around frames 30, 80 and 90 maybe attributed to the lens halo effect interacting with the target and can be clearly seen in the observed images.

In SEQ2, the target moves outside the sensors view range for a few frames and re-enters the scene. By frame 195 the target is only partially visible and this is reflected in the increased uncertainty around those frames in Fig. 12b. A subsequent exit from the scene triggers the detector mode of the $AKF_{als+tld}$ tracker. The detector is able to quickly detect the target upon re-entry and passes it on to the tracker when a uncertainty value below the threshold is achieved. The AKF_{als} tracker slowly deviates from the target and begins to concentrate on the background as shown in Fig. 13b.

In SEQ3, the target becomes absent from the scene for an extended period of time before re-entering. The exit of the target around frame 263 is easily picked up by the uncertainty indicator. Note the gradual increase in uncertainty corresponding to the slow exit of the target from the scene. When the target re-enters the scene, the detector is quick to move on to the true target, however, the uncertainty associated with it still remains high as seen in Figs. 11 and 12c. This suggests that, though the uncertainty criterion is robust enough to detect track-losses, it may not be a strong indicator of track-acquisition. In cases of long absences it may be necessary to re-initialize the tracker with a more robust detector using feature descriptors that are more complex than intensity histograms.

8 Conclusion

We have discussed infrared tracking under a unified co-inference framework where both Kalman filtering and particle filtering are involved to update target appearance and to estimate target kinematics, sequentially and respectively. We have proposed a dual foreground–background appearance model that integrates local statistics of both background and foreground to enhance the tracker's sensitivity and robustness. The key to this research is how to robustly and reliably update the target appearance represented by multiple histograms during the tracking process. Particularly, we have proposed a new AKF-based appearance learning method, AKF_{als}, which is compared with the two existing techniques, AKF_{cov} and HS. It is shown that AKF_{als} demonstrates the best performance and also supports robust occlusion handling.

Acknowledgments The authors thank Dr. James B. Rawlings's research group at the University of Wisconsin-Madison for providing the ALS code.[3] This work was supported in part by the U.S. Army Research Laboratory and the U.S. Army Research Office under Grants W911NF-04-1-0221 and W911NF-08-1-0293 and the 2009 Oklahoma NASA EPSCoR Research Initiation Grant (RIG).

References

1. Arulampalam, M.S., Maskell, S., Gordon, N., Clapp, T.: A tutorial on particle filters for online nonlinear/non-Gaussian Bayesian tracking. IEEE Trans. Signal Process. **50**(2), 174–188 (2002)
2. Birchfield, S.: Elliptical head tracking using intensity gradients and color histograms. In: Proceedings of the IEEE International Conference on Computer Vision and Pattern Recognition, pp. 232–237 (1998)
3. Carew, B., Belanger, P.: Identification of optimum filter steady-state gain for systems with unknown noise covariances. IEEE Trans. Autom. Control **18**(6), 582–587 (1973)
4. Collins, R.T., Liu, Y., Leordeanu, M.: Online selection of discriminative tracking features. IEEE Trans. Pattern Anal. Mach. Intell. **27**(10), 1631–1643 (2005)
5. Comaniciu, D., Ramesh, V., Meer, P.: Real-time tracking of non-rigid objects using mean shift. In: Proceedings of the IEEE International Conference on Computer Vision and Pattern Recognition, vol. 2, pp. 142–149 (2000)
6. Comaniciu, D., Ramesh, V., Meer, P.: Kernel-based object tracking. IEEE Trans. Pattern Anal. Mach. Intell. **25**(5), 564–577 (2003)
7. Dawoud, A., Alam, M.S., Bal, A., Loo, C.: Decision fusion algorithm for target tracking in infrared imagery. Opt. Eng. **44**, 026401–18 (2005)
8. Dawoud, A., Alam, M.S., Bal, A., Loo, C.: Target tracking in infrared imagery using weighted composite reference function-based decision fusion. IEEE Trans. Image Process. **15**(2), 404–410 (2006)
9. del Blanco, C.R., Jaureguizar, F., Garcìa, N., Salgado, L.: Robust automatic target tracking based on a Bayesian ego-motion compensation framework for airborne FLIR imagery. In: Sadjadi, F.A., Mahalanobis, A. (eds.) Polarimetric and Infrared Processing for ATR. Proceedings of the SPIE, vol. 7335 (2009)
10. Han, B., Zhu., Y., Comaniciu, D., Davis, L.: Kernel-based Bayesian filtering for object tracking. In: Proceedings of the IEEE International Conference on Computer Vision and Pattern Recognition, vol. 1, pp. 227–234 (2005)
11. Han, T.X., Liu, M., Huang, T.S.: A drifting-proof framework for tracking and online appearance learning. In: Proceedings of the IEEE Workshop on Applications of Computer Vision (2007)
12. Harger, G.D., Belhumeur, P.N.: Real-time tracking of image regions with changes in geometry and illumination. In: Proceedings of the IEEE International Conference on Computer Vision and Pattern Recognition, San Francisco, CA, pp. 403–410 (1996)
13. Jepson, A.D., Fleet, D.J., El-Maraghi, T.F.: Robust online appearance models for visual tracking. IEEE Trans. Pattern Anal. Mach. Intell. **25**(10), 1296–1311 (2003)
14. Johnston, C.M., Mould, N., Havlicek, J.P., Fan, G.: Dual domain auxiliary particle filter with integrated target signature update. In: Proceedings of the IEEE International Conference on Computer Vision and Pattern Recognition Workshops, Miami, FL, pp. 54–59 (2009)
15. Khan, J.F., Alam, M.S.: Efficient target detection in cluttered FLIR imagery. In: Casasent, D.P., Chao, T.-H. (eds.) Optical Pattern Recognition XVI. Proceedings of the SPIE, vol. 5816, pp. 39–53 (2005)

[3] http://jbrwww.che.wisc.edu/software/als/

16. Lankton, S., Malcolm, J., Nakhmani, M.A., Tannenbaum, A.: Tracking through changes in scale. In: Proceedings of the International Conference on Image Processing, pp. 241–244 (2008)
17. Latecki, L.J., Miezianko, R.: Object tracking with dynamic template update and occlusion detection. In: Proceedings of the International Conference on Pattern Recognition, Hong Kong, China, vol. 1, pp. 556–560 (2006)
18. Leichter, I., Lindenbaum, M., Rivlin, E.: Tracking by affine kernel transformations using color and boundary cues. IEEE Trans. Pattern Anal. Mach. Intell. **31**(1), 164–171 (2009)
19. Li, X., Hu, W.M., Zhang, Z.F., Zhang, X.Q., Zhu, M.L., Cheng, J.: Visual tracking via incremental log-Euclidean Riemannian subspace learning. In: Proceedings of the IEEE International Conference on Computer Vision and Pattern Recognition, pp. 1–8 (2008)
20. Li, X.R., Bar-Shalom, Y.: A recursive multiple model approach to noise identification. IEEE Trans. Aerosp. Electron. Syst. **30**(3), 671–684 (1994)
21. Maggio, E., Cavallaro, A.: Hybrid particle filter and mean shift tracker with adaptive transition model. In: Proceedings of the IEEE International Conference on Acoustics, Speech, and Signal Processing, vol. 2, pp. 221–224 (2005)
22. Matthews, L., Ishikawa, T., Baker, S.: The template update problem. IEEE Trans. Pattern Anal. Mach. Intell. **26**(6), 810–815 (2004)
23. Maybeck, P.S., Rogers, S.K.: Adaptive tracking of multiple hot-spot target IR images. IEEE Trans. Autom. Control **28**(10), 937–943 (1983)
24. Mehra, R.K.: On the identification of variances and adaptive Kalman filtering. IEEE Trans. Autom. Control **15**(2), 175–184 (1970)
25. Mehra, R.K.: Approaches to adaptive filtering. IEEE Trans. Autom. Control **17**(5), 693–698 (1972)
26. Mohamed, A.H., Schwarz, K.P.: Adaptive Kalman filtering for INS/GPS. J. Geodesy **73**, 193–203 (1999)
27. Mould, N.A., Nguyen, C.T., Johnston, C.M., Havlicek, J.P.: Online consistency checking for AM-FM target tracks. In: Bouman, C.A., Miller, E.L., Pollak, I. (eds.) Proceedings of the SPIE/IS&T Conference on Computational Imaging VI. Proceedings of the SPIE, vol. 6814 (2008)
28. Neethling, C., Young, P.: Comments on "identification of optimum filter steady-state gain for systems with unknown noise covariances". IEEE Trans. Autom. Control **19**(5), 623–625 (1974)
29. Nguyen, H.T., Smeulders, A.W.M.: Fast occluded object tracking by a robust appearance filter. IEEE Trans. Pattern Anal. Mach. Intell. **26**(8), 1099–1104 (2004)
30. Nguyen, H.T., Worring, M., van der Boomagaard, R.: Occlusion robust adaptive template tracking. In: Proceedings of the IEEE International Conference on Computer Vision, Vancouver, BC, vol. 1, pp. 678–683 (2001)
31. Noriega, G., Pasupathy, S.: Adaptive estimation of noise covariance matrices in real-time preprocessing of geophysical data. IEEE Trans. Geosci. Rem. Sens. **35**(5), 1146–1159 (1997)
32. Odelson, B.J., Lutz, A., Rawlings, J.B.: The autocovariance least-squares method for estimating covariances: application to model-based control of chemical reactors. IEEE Trans. Control Syst. Technol. **14**(3), 532–540 (2006)
33. Odelson, B.J., Rajamani, R.M., Rawlings, B.J.: A new autocovariance least-squares method for estimating noise covariances. Automatica **42**(2), 303–308 (2006)
34. Pan, J., Hu, B.: Robust object tracking against template drift. In: IEEE International Conference on Image Processing, pp. 353–356 (2007)
35. Peng, N.S., Yang, J., Liu, Z.: Mean shift blob tracking with kernel histogram filtering and hypothesis testing. Pattern Recognition Lett. **26**(5), 605–614 (2005)
36. Peng, Z., Zhang, Q., Guan, A.: Extended target tracking using projection curves and matching pel count. Opt. Eng. **46**(6), 0664011–0664016 (2007)
37. Porikli, F., Tuzel, O., Meer, P.: Covariance tracking using model update based on Lie algebra. In: Proceedings of the IEEE International Conference on Computer Vision and Pattern Recognition, vol. 1, pp. 728–735 (2006)

38. Powers, R.M., Pao, L.Y.: Using Kolmogorov–Smirnov tests to detect track-loss in the absence of truth data. In: Proceedings of the IEEE Conference on Decision and Control, pp. 3097–3104 (2005)
39. Shaik, J.S., Iftekharuddin, K.M.: Automated tracking and classification of infrared images. In: Proceedings of the International Joint Conference on Neural Networks, vol. 2, pp. 1201–1206 (2003)
40. She, K., Bebis, G., Gu, H., Miller, R.: Vehicle tracking using on-line fusion of color and shape features. In: Proceedings of the IEEE International Conference on Intelligent Transportation Systems, pp. 731–736 (2004)
41. Swain, M.J., Ballard, D.H.: Indexing via color histograms. In: Proceedings of the International Conference on Computer Vision, pp. 390–393 (1990)
42. Tuzel, O., Porikli, F., Meer, P.: Region covariance: a fast descriptor for detection and classification. In: Proceedings of the 9th European Conference on Computer Vision, pp. 589–600 (2006)
43. Venkataraman, V., Fan, G., Fan, X.: Target tracking with online feature selection in flir imagery. In: Workshop on Object Tracking and Classification Beyond the Visible Spectrum, pp. 1–8 (2007)
44. Wang, Z., Wu, Y., Wang, J., Lu, H.: Target tracking in infrared image sequences using diverse adaboostsvm. In: Proceedings of the International Conference on Innovative Computing, Information and Control, USA, pp. 233–236. IEEE Computer Society, Washington, DC (2006)
45. Wu, Y., Huang, T.S.: Robust visual tracking by integrating multiple cues based on co-inference learning. Int. J. Comput. Vis. **58**(1), 55–71 (2004)
46. Wu, Y., Yu, T., Hua, G.: Tracking appearances with occlusions. In: Proceedings of the Computer Vision and Pattern Recognition, vol. 1, pp. 789–795 (2003)
47. Yi, S., Zhang, L.: A novel multiple tracking system for UAV platforms. In: Henry, D.J. (ed.) ISR Systems and Applications III. Proceedings of the SPIE, vol. 6209 (2006)
48. Yilmaz, A., Li, X., Shah, M.: Contour-based object tracking with occlusion handling in video acquired using mobile cameras. IEEE Trans. Pattern Anal. Mach. Intell. **26**(11), 1531–1536 (2004)
49. Yilmaz , A., Shafique, K., Shah, M.: Tracking in airborne forward looking infrared imagery. Image Vis. Comput. **21**(7), 623–635 (2003)
50. Yilmaz, A., Javed, O., Shah, M.: Object tracking: a survey. ACM Comput. Surv. **38**(4), 13 (2006)
51. Zhang, C., Rui, Y.: Robust visual tracking via pixel classification and integration. In: Proceedings of the International Conference on Pattern Recognition, vol. 3, pp. 37–42 (2006)
52. Zivkovic, Z., Krose, B.: An EM like algorithm for color-histogram-based object tracking. In: Proceedings of the IEEE International Conference on Computer Vision and Pattern Recognition, vol. 1, pp. 798–803 (2004)

3D Model-Driven Vehicle Matching and Recognition

Tingbo Hou, Sen Wang and Hong Qin

Abstract Matching vehicles subject to both large pose transformations and extreme illumination variations remains a technically challenging problem in computer vision. In this chapter, we first investigate the state-of-the-art studies on vehicle matching, inverse rendering by which illumination can be factorized from the light reflectance field, and applications of the near-IR illumination in computer vision. Then a 3D model-driven framework is developed, towards matching and recognizing vehicles with varying pose and (visible or near-IR) illumination conditions. We adopt a compact set of 3D models to represent basic types of vehicle. The pose transformation is estimated by using approximated vehicle models that can effectively match objects under large viewpoint changes and partial occlusions. With the estimation of surface reflectance property, illumination conditions are approximated by a low-dimensional linear subspace using spherical harmonics representation. By estimated pose and illumination conditions, we can re-render vehicles in the reference image to generate the relit image with the same pose and illumination conditions as the target image. Finally, we compare the relit image and the re-rendered target image to match vehicles in the original reference image and target image. Furthermore, no training is needed in our framework and re-rendered vehicle images in any other viewpoints and illumination conditions can be obtained from just one single input image. In our experiments, both synthetic data and real data are used. Experimental

T. Hou (✉) · H. Qin
Department of Computer Science, Stony Brook University (SUNY),
Stony Brook, NY 11794, USA
e-mail: thou@cs.sunysb.edu

H. Qin
e-mail: qin@cs.sunysb.edu

S. Wang
Kodak Research Laboratories, Eastman Kodak Company, Rochester,
NY 14650, USA
e-mail: sen.wang@kodak.com

R. Hammoud et al. (eds.), *Machine Vision Beyond Visible Spectrum*,
Augmented Vision and Reality, 1, DOI: 10.1007/978-3-642-11568-4_3,

results demonstrate the robustness and efficacy of our framework, with a potential to generalize our current method from vehicles to handle other types of objects.

Keywords Vehicle matching · 3D model-driven method · Inverse rendering · Spherical harmonics · Near-IR illumination

1 Introduction

Object matching and recognition remain an important and long-term task with continuing interest from computer vision and various applications in security, surveillance, and robotics. Many types of representations have been exploited to match and recognize objects by a set of low-dimensional parameters, such as shape, texture, structure, and other specific feature patterns. However, when it comes to unconstrained conditions such as highly varying pose and severely changing illumination, the problem becomes extremely challenging. As shown in Fig. 1, object appearance may be tremendously different with varying pose and illumination conditions. Although the texture of a vehicle is consistent, its appearance indeed varies a lot under different lightings. Thus, such clues like shape and texture are weak in this case.

Currently, popular approaches in object recognition focus on two trends: the appearance-based methods [4, 16] and the model-based methods [7, 20]. In appearance-based methods, objects are typically represented by a group of feature vectors, and a set of positive and negative examples is adopted to train a classifier spanning on the principle component analysis (PCA) subspace or feature subspace. In practice, technical issues arise from appearance variation due to different pose and lightings. Model-based methods require a set of 3D models to provide geometric constraints. Ideally, when object domain is known, the explicit utilization of 3D models can largely alleviate the problem of feature matching. However, it stands on two basic assumptions: first, the 3D model can precisely fit to the input images; second, pose estimation is accurate enough. To estimate appearance of objects, global and local clues have been used to simulate texture of the 3D model. Despite the progress, it still has limited success in illumination variations, since illumination conditions can dramatically affect appearances as shown in Fig. 1.

The union of model and illumination is appealing, since appearances can be decomposed and reassembled by them. It provides a pose and illumination invariant view to examine the problem of matching and recognition. However for general objects, it is hard to obtain their 3D models from single image. To alleviate this restriction, we choose vehicle as object domain with simple geometric structure. The illumination can be visible spectrum or near-Infrared (IR) spectrum. The near-IR light can also be reflected by objects since it is close to the visible light in spectrum. One benefit of near-IR illumination is to allow our method work in low-luminance environment with active near-IR light sources. Thus the primary contribution of this chapter lies in a 3D model-driven framework towards vehicle matching and recognition working under visible or near-IR illumination, which can handle large pose

Fig. 1 Images of the same vehicle taken from different viewpoints and lightings, subject to large pose and illumination variations

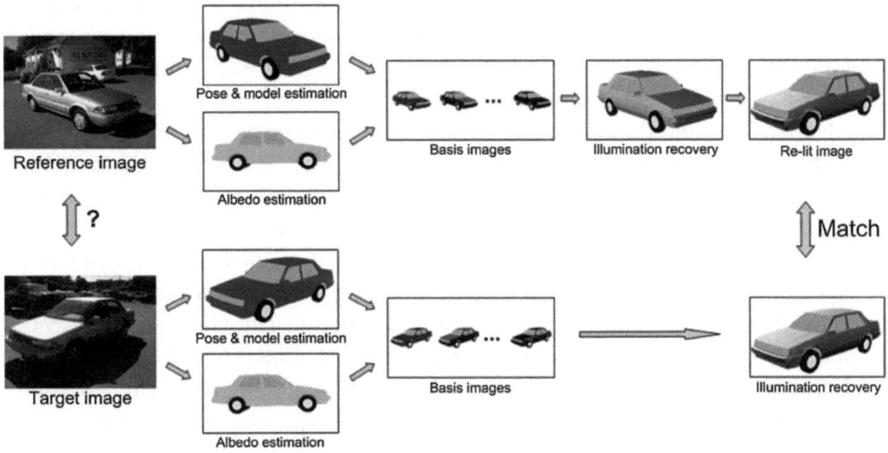

Fig. 2 Our vehicle matching framework. Given input images, the model and pose transformation are first determined. Then we estimate reflectance property (albedo) of objects. The illumination is recovered by spanning given image to basis images, and the re-lit image is generated by transferring illumination. Finally, objects are compared in a shared domain with recovered illumination

transformations and illumination variations simultaneously. Our vehicle matching framework is shown in Fig. 2. Given original input images, the pose transformation is first estimated by using approximated 3D vehicle models that can effectively match objects under large viewpoint changes and partial occlusions. Second, we estimate reflectance property of objects, taking advantage of the fact that the body of a vehicle has unified color and material. After that, we compute their spherical harmonic basis and recover illumination conditions both in the reference image and target image. By effectively estimating both pose and illumination conditions, we can re-render vehicles in the reference image to generate the relit image with the same pose and illumination conditions as the target image. Finally, we make comparisons between the relit image and the re-rendered target image to match vehicles in the original reference image and target image.

2 Previous Work

In this section, we will investigate previous related work in vehicle matching, inverse rendering and near-IR illumination.

2.1 Vehicle Matching

Vehicle matching has been studied in many areas of computer vision, with different purposes such as detection, identification, tracking, and recognition. Appearance-based methods are well applied on vehicles, with no difference with other objects. Recently, Shan et al. [22] exploited an embedding vector to represent each vehicle image by exemplars of vehicles within the same camera. Each component of this vector is a non-metric distance computed by oriented edge maps. The measurement they defined describes the appearance-based same-different probabilities of two vehicles. The extended work was done by Guo et al. [9] and Shan et al. [23] for vehicle matching. Here, we pay more attention to model-based methods, since 3D model can connect appearances from multiple views. Thus large pose variation can be easily handled. A vehicle has concise shape that can be easily represented by a simple 3D model. Koller et al. [13] represented vehicles by a general 3D model parameterized by 12 length parameters. Their method needs to calibrate a moving plane from video sequences. Kim and Malik [12] used a simple sedan model to detect vehicle, and used probabilistic feature grouping for vehicle tracking. Guo et al. [8] proposed a model-based approach to match vehicles. They used approximate 3D models to handle pose transformation, and a piecewise Markov Random Field (MRF) model to guess texture of occluded parts. However, their method has limitations on sensitive model fitting and varying illumination. Another benefit for model-based methods is that illumination as a higher dimension can be properly analyzed when the 3D shape is known. In the work of Hou et al. [10], a vehicle matching framework was proposed using a compact set of vehicle models and spherical harmonics representation of illumination.

2.2 Inverse Rendering

Illumination can be interpreted as one of the attributes of light reflectance field. Its analysis and manipulation can be fulfilled by factorizing illumination from images, which is named "inverse rendering". Inverse rendering which measures rendering attributes: lighting, texture, and bidirectional reflectance distribution function (BRDF) from photographs, continues to be an active research area with interest from both computer vision and computer graphics. In previous work, tremendous progress has been made in the recovery of these three rendering attributes with one or two unknowns [3, 21, 27]. In general cases where lighting, texture, and BRDF are all unknown, this problem becomes ill-conditioned until strong assumptions and requirements on input data have been made. Ramamoorthi and Hanrahan [19] presented a signal processing framework for inverse rendering with known geometry and isotropic BRDFs. In their work, the reflected light field was expressed as a convolution of the lighting and BRDF using spherical harmonics. As a frequency-space convolution, spherical harmonics has been used as a tool to represent lighting. In the work of [1], it is shown that the reflected light field from a Lambertian surface can be characterized using only its first nine spherical harmonic coefficients, where geometry is assumed to be known. Later, Zhang et al. [28, 29] integrated the spherical harmonic illumination representation into the Morphable Model approach, by modulating the texture component with the spherical harmonic bases. They used PCA to initialize geometry and texture from a large set of training data, and estimate lighting and basis images independently through iteration. To alleviate the strong requirements on geometry and texture, Wang et al. [25] proposed a subregion based framework that uses a MRF to model the statistical distribution and spatial coherence of texture. Though lighting in a small region is more homogeneous, it is hard to segment an image into homogeneous regions. Their method still needs training data to compute PCA texture model.

2.3 Near-IR Illumination

Low-cost infrared cameras make it possible to address computer vision problems in a larger range of the electromagnetic spectrum [15]. Here, we only focus on the near-infrared illumination with wavelength varying from 0.7 to 1 μm. It is very close to the visible spectrum, and thus it can be reflected by objects, generating IR images similar with images under visible spectrum. Novotny and Ferrier [17] used active IR to measure distance. They proposed a method of determining the reflectance property of a surface under infrared illumination using Phong model, since IR LEDs are well approximated as a point light source. Ji and Yang [11] studied real time 3D face pose discrimination based on active IR illumination. The IR is adopted since pupils in IR images are more clear and stable than images under visible illumination. In the work of Zhu et al. [31], Zhao and Grigat [30], active near-IR illumination was employed in eye detection and eye tracking. Zou et al. [32] used active near-IR illumination projected by LED light source to illumination invariant face recognition. The near-IR light source can provide constant illumination, and produce images with higher quality than images under ambient illumination. More work on face recognition using active near-IR illumination can be found in the work of Pan et al. [14, 18]. Wang et al. [24] presented a method for relighting faces for reducing the effects of uneven lighting and color in video conference. Their setup consists of a compact lighting rig and two cameras. The IR camera is 8 times (120 fps/15 fps) faster than the color camera. They used active IR lights to obtain an illumination bases of the scene, and thus they can image relighting. In the work of Fredembach and Süsstrunk

Fig. 3 Some 3D vehicle models adopted in our framework. Models are selected from the Princeton Shape Benchmark

[6], illuminant was detected and estimated in near-IR images by simply looking at the ratios of two images: a standard RGB image and a near-IR only image. As the differences between illuminants are amplified in the near-infrared, this estimation proves to be more reliable than using only the visible band.

3 Vehicle Matching Framework

In this section, we will introduce our framework of vehicle matching under various pose and illumination conditions.

3.1 Model Determination

Our dataset contains five representative models that stand for five different categories of vehicles including compact-size car, full-size sedan, small pickup truck, SUV, and big truck, respectively. These 3D models are selected from the Princeton Shape Benchmark,[1] with some of them shown in Fig. 3. Unlike the approach by Guo et al. [8], which requires each vertex in 3D model has its semantic ownership, we take the body of a vehicle as an object and ignore some parts (windows, wheels, and lights) for such reasons: (1) typically, the body has uniform color and material, which leads to uniform reflectance property to illumination; (2) the removed parts have different patterns and properties. For example, windows could have mirror reflection, wheels may turn right or left with the same pose of the body, and lights could be on or off.

For each input image, we will first determine which model best represents the vehicle that appears in the image. Considering the fact that pose estimation is easily trapped into local minimum in the searching space, we select three different initial poses for each vehicle model with reasonable projection. For each model, we compute edge maps under these three initial fittings and use chamfer distance [22] to measure the similarity with edge maps of the original input image, as shown in Fig. 5. Finally, we select the top two matched models as candidates for the next step.

1 (http://shape.cs.princeton.edu/benchmark/)

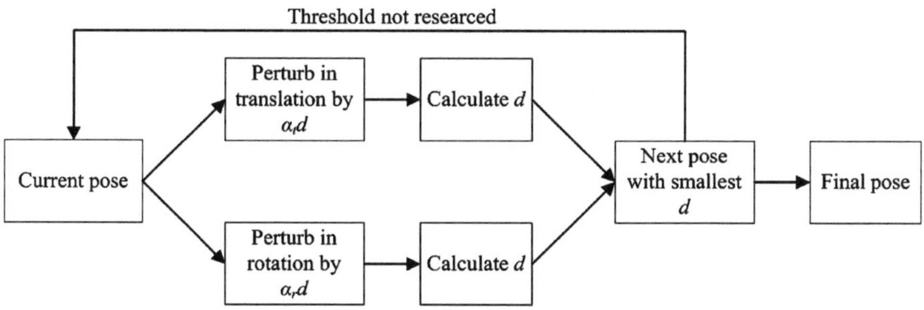

Fig. 4 Flow chart of pose recovery. Refer to texts for more details

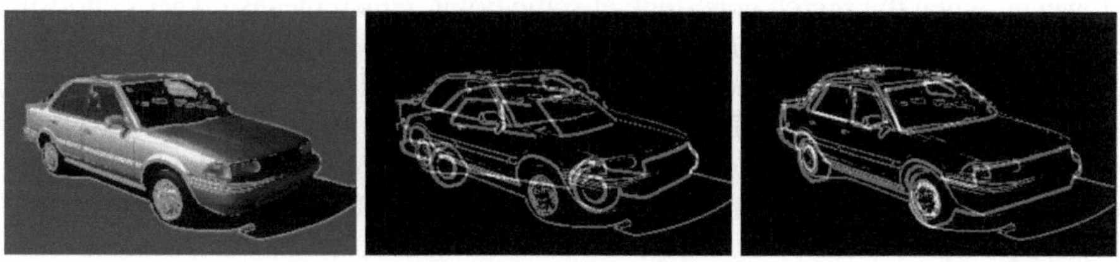

Fig. 5 Pose estimation. Edge detection in the original image is shown on the *left*. Initial fitting between 3D vehicle model and the original image is shown in the *middle*. The alignment is shown on the *right*.

3.2 Pose Recovery

Here, pose recovery refers to aligning a 3D model to an object in the image. This task is easier to perform if the 3D model and the image have similar features. A few correspondences will be enough to perform the alignment, since the objects are rigid. However, the visual contents of geometric models are unknown at this stage. We only have their geometric information, i.e., 3D coordinates and normals. So it gives rise to a simple question: How to compare geometry with texture?

An intuitive idea is to utilize the geometric edges, i.e., silhouettes and intersecting lines of two smooth surfaces, which happen to appear in the edge maps of images. We employ an approach inspired by the one in [8] that used an Iterative Closet Point (ICP)-directed search to iteratively align the geometric edges to the image edges. The searching space is spanned by six independent components: three translation elements and three rotation angles along three axes, by isometric sampling. By making this simplification, we assume that the intrinsic parameters of cameras are fixed.

The flow chart of our pose recovery is shown in Fig. 4. For a current pose of a candidate model, we search for the next best pose with minimal average closet point distance d, given by

$$d = \frac{1}{N}\sum_{i=1}^{N} d_i, \tag{1}$$

where d_i denotes the distance between pixel i in the geometric edges and its closet pixels in the image edges. We sample the searching space of translation and rotation respectively by perturbing a sampling distance. In 3D translation, we use three samplings for each direction, with positive, zero, and negative distance, that is, 27 samplings, and similarly in rotation, 27 samplings in three angles along three axes. The scale of geometric edges can be adjusted through the translation in the direction of depth. Furthermore, we employ adaptive sampling distances $\alpha_t d$ and $\alpha_r d$, where α_t and α_r are scaling parameters of translation and rotation. Thus, the search speed can be controlled in the way that when it is getting closer to the minimum distance, the sampling distance is getting smaller to achieve a more precise search.

The search will stop when the average closest point distance d reaches a threshold. However, when the search gets stuck at some point, which means it keeps choosing zero sampling distance, while the threshold has not been reached, the sampling distance will jump to Ds, where D is a large factor to pull the search out of the local minimum. Finally, the best 3D object model is selected with minimal average closest point distance from candidate models. Figure 5 shows an example of pose recovery, where image edges and geometry edges of the 3D vehicle model are aligned.

3.3 Estimation of Reflectance Fraction

Albedo is the fraction of light that a surface point reflects when it is illuminated, which is an intrinsic property that depends on materials of the surface. There are some approaches in literature to estimate albedo from a single image [2]. In previous work of applying spherical harmonics [29], the brightness of a pixel is taken as albedo. In our framework, taking the observation that the body of a vehicle has uniform texture and materials, we estimate albedo in RGB channels for visible spectrum.

For Lambertian objects, the diffused component of the surface reflection satisfies Lambertian Cosine Law (as shown in Fig. 6), given by

$$I = \rho \max(n^T s, 0), \tag{2}$$

where I is the pixel intensity, s is the light source direction, ρ is the reflectance fraction of the surface (albedo), and n is the surface normal of the corresponding 3D points. The expression implicitly assumes a single dominant light source placed at infinity, which is the most common case where vehicle images are taken. Note that Lambertian law in its pure form is nonlinear due to the max function, which accounts for the formation of attached shadows. Shadows and specularities do not reveal any information about their reflectivity. Thus they should not be included in the computation of estimation. In most cases, vehicle images are taken in an outdoor environment where the primary light source is the sun, and thus the estimation is realistic.

By collecting 3D points with positive $(n^T s)$ and the corresponding image pixels excluding shadows and specularities, we can obtain a reflective equation for each

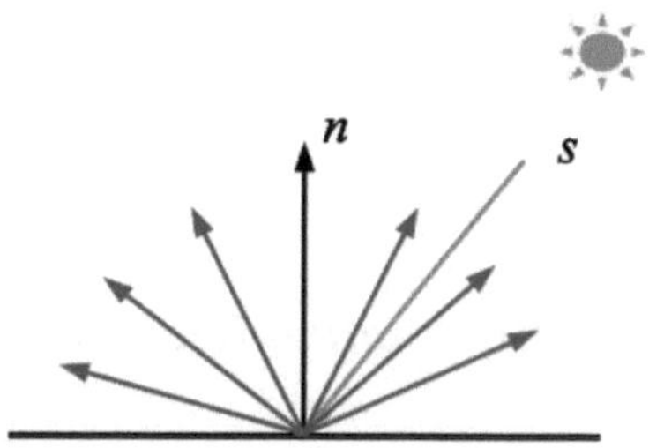

Fig. 6 Lambertian reflectance with point light source: light is scattered equally to every direction regardless of the observer's angle of view, and the intensity is subject to the Lambertian Cosine Law

Fig. 7 Reflectance fraction (albedo) estimation. The input images are shown in the *first row* and estimated albedos are shown in the *second row*. The *two left images* come from the same car and the *right-most image* comes from another car

point in the 3D model, written as:

$$n^T \rho s = I. \tag{3}$$

Note that s is almost the same for each point in the 3D model, since the only dominant light source is placed at infinity. Therefore, we can get a formula for all reflective equations as (for example in the red channel):

$$N\rho_r s = I_r, \tag{4}$$

where N is the $n \times 3$ matrix that consists of surface normals of n points, ρ_r is the albedo in the red channel, and I_r is intensity value of the red component of n corresponding pixels in the image. And so are the green and blue channels. We then take $\rho_r s$ together as a variable and estimate it by the method of least squares. Since ρ_r is a positive fraction in the range [0, 1], and s is the normalized direction vector whose length equals 1, we can compute ρ_r by

$$\rho_r = \frac{|\rho_r s|}{|s|} = |\rho_r s|. \tag{5}$$

Similarly, we can compute albedo in green channel ρ_g and albedo in blue channel ρ_b. Figure 7 shows that albedo maps in the second row are estimated from three input images in the first row. The two left images are taken from the same car and the right-most image is from another car. Despite varying illuminations, the albedo estimation is accurate and robust.

3.4 Illumination Recovery

As described by Basri and Jacobs [1] and Ramamoorthi and Hanrahan [19]), any image under arbitrary illumination conditions can be approximately represented by a linear combination of spherical harmonic basis as:

$$I \approx bl, \tag{6}$$

where b is the spherical harmonic basis and l is the vector of illumination coefficients. The set of images of a convex Lambertian object obtained under a wide variety of lighting conditions can be approximated accurately by a 9-dimensional linear subspace [1, 19, 28]. They are the spherical analog of the Fourier basis on the line or circle. The first nine spherical harmonic basis images of an object can be computed by:

$$\begin{aligned}
&b_{00} = \frac{1}{\sqrt{4\pi}}\lambda, && b_{10}^{e} = \sqrt{\frac{3}{4\pi}}\lambda .* n_z, \\
&b_{11}^{o} = \sqrt{\frac{3}{4\pi}}\lambda .* n_y, && b_{11}^{e} = \sqrt{\frac{3}{4\pi}}\lambda .* n_x, \\
&b_{20} = \frac{1}{2}\sqrt{\frac{3}{4\pi}}\lambda .* (2n_{z^2} - n_{x^2} - n_{y^2}), && \\
&b_{21}^{o} = 3\sqrt{\frac{5}{12\pi}}\lambda .* n_{yz}, && b_{21}^{e} = 3\sqrt{\frac{5}{12\pi}}\lambda .* n_{xz}, \\
&b_{22}^{o} = 3\sqrt{\frac{5}{12\pi}}\lambda .* n_{xy}, && b_{22}^{e} = \frac{3}{2}\sqrt{\frac{5}{12\pi}}\lambda .* (n_{x^2} - n_{y^2}),
\end{aligned} \tag{7}$$

where the superscripts e and o denote the odd and the even components of the harmonics, respectively, λ is the vector of the object's albedo, n_x, n_y, n_z are three vectors of the same length that contain the x, y, and z components of the surface normals. Furthermore, n_{xy} is a vector such that the ith element $n_{xy,i} = n_{x,i}n_{y,i}$, and $\lambda .* v$ denote the component-wise product of λ with any vector v.

In our framework, we use unified estimated albedo for the body of the vehicle model. The visible part of a 3D vehicle model, which is projected to the input image due to recovered pose, provides us normal vectors, estimated albedo, and appearances with illumination effects for each visible 3D point associated with corresponding 2D pixels. Therefore, we can compute the first nine spherical harmonic basis using Eq. 7, and estimate the illumination coefficients l by using the method of least squares in Eq. 6. Figure 8 shows an example of the first nine spherical harmonic basis images with RGB channels where light intensities represent positive values and dark intensities represent negative values.

Fig. 8 An example of the first nine spherical harmonic basis images with RGB channels. *Light intensities* represent positive values and *dark intensities* represent negative values

3.5 Re-lighting

Re-lighting is used to generate new images of the object from the reference image by transferring illumination effects in the target images [25, 26, 29]. In our framework, we use this technique to render the reference object under illumination conditions of the target image. In the work of Wang et al. [24], re-lighting was constructed on basis images obtained under various active IR illumination. Our basis images are from the spherical harmonic bases. By Eq. 6, we obtain two illumination representations of both the reference image and the target image:

$$I_r \approx b_r l_r, \quad I_t \approx b_t l_t, \tag{8}$$

where the subscript r denotes the reference object, and subscript t denotes the target object. By re-lighting, we can transfer the illumination effects from the target image to the reference object if they are subject to the same pose:

$$I_{\text{relit}} \approx b_r l_t, \tag{9}$$

where I_{relit} is the relit images of the reference object with the illumination conditions of the target image.

With this re-lighting technique, we can render an object under any pose and illumination conditions associated with one single input image. Figure 9 shows examples of illumination recovery and re-lighting. From the results, we can see that the relit image (Fig. 9c) is very similar to the re-rendered image (Fig. 9e), and the relit image (Fig. 9f) is very similar to the re-rendered image (Fig. 9b). Therefore, we just compare the relit image with the re-rendered target image to match vehicles in the original reference image and target image in spite of large variations of pose and illumination.

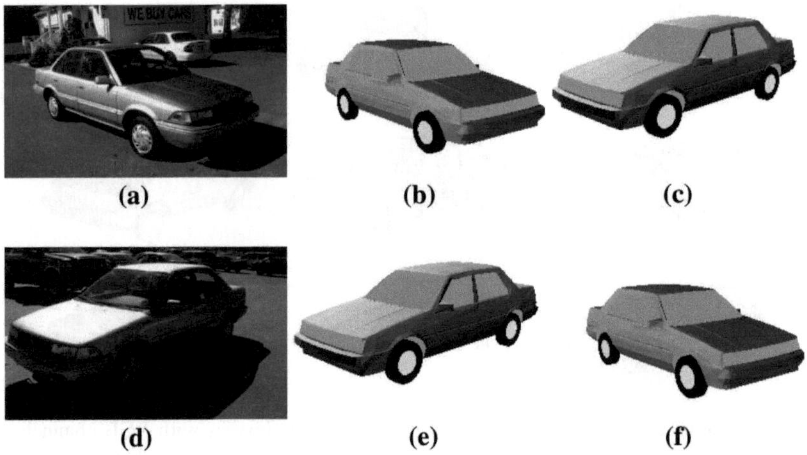

Fig. 9 Examples of illumination recovery and re-lighting. **a** and **d** are two input images. **b** and **e** are re-rendered images of **(a)** and **(d)** after illumination recovery. **c** is the relit image by transferring both pose information and illumination effects from **(d)** to the 3D vehicle model estimated from **(a)**. **f** is the relit image by transferring both pose information and illumination effects from **(a)** to the 3D vehicle model estimated from **(d)**. The relit images **(c)** and **(f)** are very similar to the re-rendered images **(e)** and **(b)**, respectively

3.6 Vehicle Matching

In order to match two images, we use the normalized matching distance (NMD), defined as

$$\mathrm{NMD} = \frac{\sum_{i=0}^{n} \| I_{\mathrm{relit}}^{i} - I_{t}^{i} \|}{\sum_{j=0}^{n} I_{t}^{j}}, \tag{10}$$

where I_{relit} is the relit image and I_t is the re-rendered image of target objects. NMD describes the difference between the reference object and the target object, in spite of the affect of pose and illumination variations. A smaller distance stands for higher similarity, and vice versa.

The vehicle matching algorithm in our framework can be summarized as follows:

1. Determine 3D vehicle models and recover their poses in both the reference image and target image.
2. Estimate reflectance fractions (albedos) from two input images by Eq. 5.
3. Compute the spherical harmonic basis and illumination coefficients for each input image, respectively, by Eqs. 6 and 7.
4. Re-render the target object by Eq. 8 and re-lighting the reference object by Eq. 9.
5. Compare the relit image and the re-rendered image by computing the normalized matching distance by Eq. 10 to match vehicles in the original reference image and target image.

Fig. 10 Near-IR and visible spectrum image pairs of the same scenes. Images are from the data in [5]

4 Near-IR Illumination

Surfaces of objects have similar reflectance property under active near-IR illumination and visible illumination. Therefore, the framework proposed above can also be applied to images under near-IR illumination. One benefit of using near-IR illumination is that it can provide constant illumination, and work in low-luminance environment without conspicuous light. Besides, specular reflection, which is not considered by our illumination model, is significantly reduced in near-IR image. The disadvantage of near-IR illumination is that it does not have any color information. It is not significant to objects like human face, but is important to vehicle, since vehicle has plentiful colors that can be discriminated easily in visible illumination. Figure 10 shows two image pairs of near-IR image and visible spectrum image of the same scenes, obtained from the work of Fredembach and Süsstrunk [6]. Near-IR images are very close to gray images under visible spectrum.

Objects may appear unnatural under IR illumination, since many materials do not have the same reflectance fraction under visible or near-IR spectrum, e.g., a surface with green color becomes brighter in near-IR illumination than in the visible spectrum. Therefore, we do not compare images across spectral bands, which means we only match images both under visible illumination, or both under Near-IR illumination. We also assume the diffuse reflectance of Lambertian surface has a constant ratio (i.e., albedo in visible spectrum) in the same spectral band. In visible

Fig. 11 An example of near-IR vehicle image and its estimated reflectance property

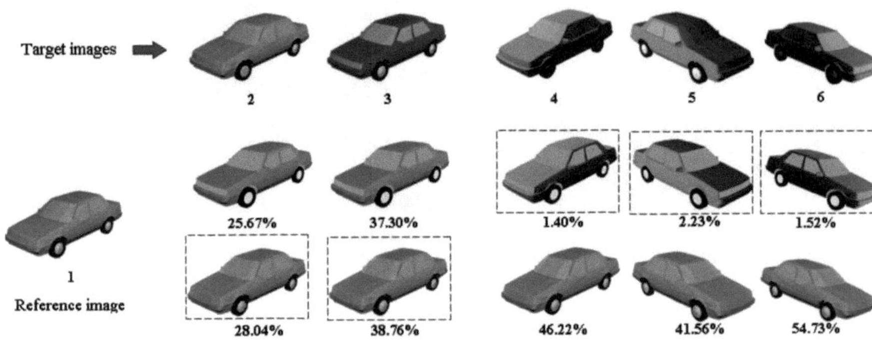

Fig. 12 Comparison on synthetic data. 1 is the reference image and 2, 3, 4, 5, 6 are target images (1, 4, 5, 6 are from the same car and 2, 3 are from another car). The *second row* are relit images by our method with their matching distances shown below. The *third row* are relit images by the method without illumination recovery by Guo et al. [8]. Our method matches the reference image with the correct target images (4, 5, 6) with lower matching distances, while the method without illumination recovery matches it with the wrong target images (2, 3)

illumination, the albedo of surface has three channels, while in near-IR illumination, the reflectance property has only one channel. Figure 11 shows an example of near-IR vehicle image and its estimated reflectance property. There is not much difference with image under visible illumination.

5 Experimental Results

In this section, we will evaluate our framework using both synthetic and real data subject to various pose and illumination conditions. The dataset contains N galleries (2–7 images in each gallery) of vehicle images. Images in the same gallery are obtained from the same vehicle under different poses and varying lightings. The evaluation scheme is to take one probe image to recognize which gallery (object) it matches with. We also compare our methods with the method without illumination recovery by Guo et al. [8] both on synthetic data and real data.

5.1 Matching Experiments

Before our recognition experiments, we conduct matching experiments on both synthetic data and real data to show how illumination conditions will affect matching and recognition results. First, we use our 3D car models to synthesize six vehicle images rendered by OpenGL with one diffuse light source and global ambient light, as shown in Fig. 12. Images 1, 2, 3 are rendered by different cars with the same pose. Images 1, 4, 5, 6 are rendered by the same car with different poses and lightings. We match image 1 (as the probe image) to the other five images. The matching performances of our method and the method without illumination recovery are shown in the second row and the third row with their matching distances, respectively. From experimental results, we can observe that our method can correctly match image 1 with target images (4, 5, 6) with lower normalized matching distances. However, the method without illumination recovery matches the reference image with wrong target images (2 and 3) due to the effect of illumination. Even in the same illumination condition, there is still a mismatch due to viewpoint variations. For example, images 1 and 5 are under the same illumination condition but taken from different viewpoints. The method without illumination recovery uses symmetry to guess the texture of vehicles. This is not correct because one side of the car is illuminated while the other side is shaded.

Figure 13 shows matching experiments on real data. Three input images are in the first row (two reference images are in the left and right, one target image is in the middle). The right image (reference image 2) and the middle image (target image) are from the same SUV, while the left image (reference image 1) is from another vehicle. The matching results of our method are shown in the second row, and the results of the method without illumination recovery are shown in the third row. According to their matching distances in experimental results, our method correctly matches the reference image 2 and the target image, while the method without illumination recovery fails due to extreme variations of pose and illumination.

5.2 Recognition Experiments

5.2.1 Synthetic Data

Our synthetic data for recognition contains nine image galleries synthesized from nine vehicle models, each of which consists of six images under large variations of pose and lighting. First, in order to test the robustness of our framework, we randomly pick nine images belonging to three different vehicles from our synthetic dataset, and

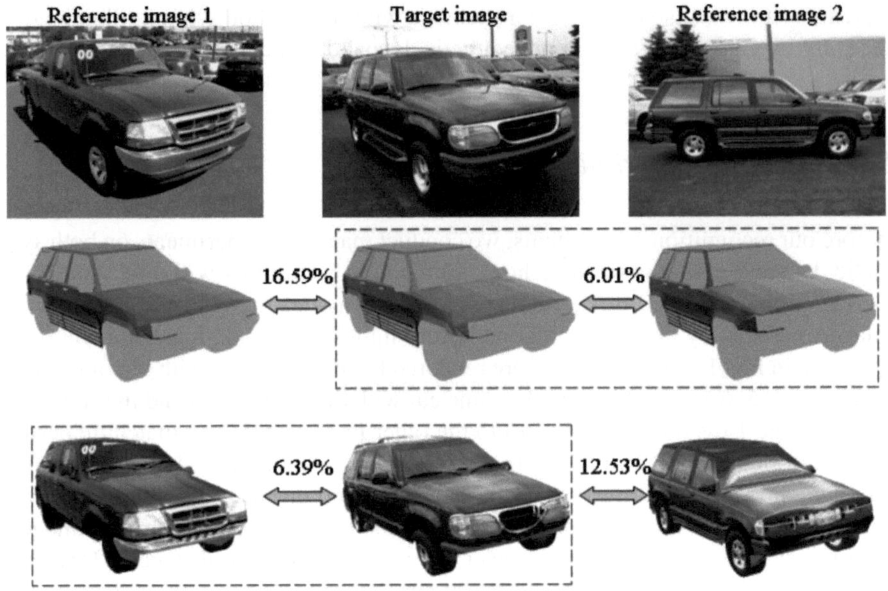

Fig. 13 Comparison on real data. Original images are shown in the *first row*. The *right image* (reference image 2) and the *middle image* (target image) are from the same SUV, while the *left image* (reference image 1) is from another vehicle. The results of our method are shown in the *second row*, and results of the method without illumination recovery are shown in the *third row*. Numbers above *arrows* connecting two images are their matching distances

compute a similarity matrix among these nine images. Figure 14 shows results by our method (left) and the method without illumination recovery (right), where the x- and y-coordinates are these nine images. We take the first image in each row as probe image and match it to the other images. Each entry of the matrix stands for a similarity between the probe images (y-coordinate) and the target images (x-coordinate). The value of each entry is illustrated by an intensity, which can be specified in the index bar: 1.0 indicates highest similarity and 0.0 indicates lowest similarity. The diagonal has similarity 1.0 where an image matches with itself. An ideal similarity matrix would have a block diagonal structure with consistently high scores on the main diagonal blocks and consistently low scores elsewhere. From results, we can see that our method provides more distinguishable bands of rows and columns between different vehicles, indicating that it has a better capability to recognize objects subject to various pose and illumination conditions. However, there is no distinct diagonal block in the right matrix, which clearly suffers from the variation of illumination.

For recognition, we conduct our recognition experiments on nine image galleries (six images in each gallery) with rank from 1 to 5 (the number of images in each gallery is from 1 to 5). Here, we test the following algorithms: (1) the method without illumination recovery by Guo et al. [8], (2) the method using average texture of vehicle

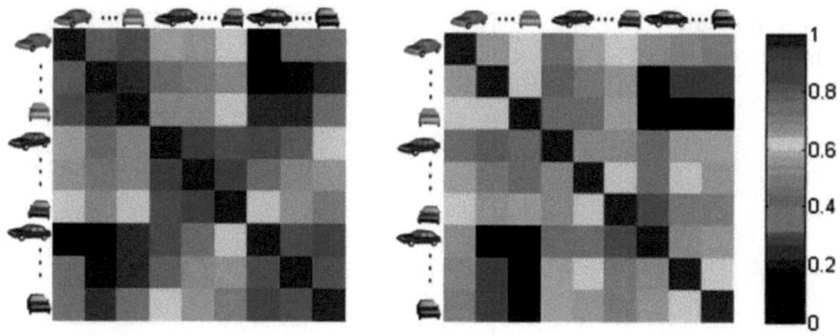

Fig. 14 Similarity matrices for our method (*left*) and the method without illumination recovery (*right*). The x and y coordinates are these nine images. The value of each entry is illustrated by an intensity which can be specified in the index bar: 1.0 indicates highest similarity and 0.0 indicates lowest similarity. The diagonal has similarity 1.0 where an image matches itself

Fig. 15 Recognition results on synthetic data: The x-coordinate denotes the size of each gallery from 1 to 5, and the y-coordinate denotes the recognition rates

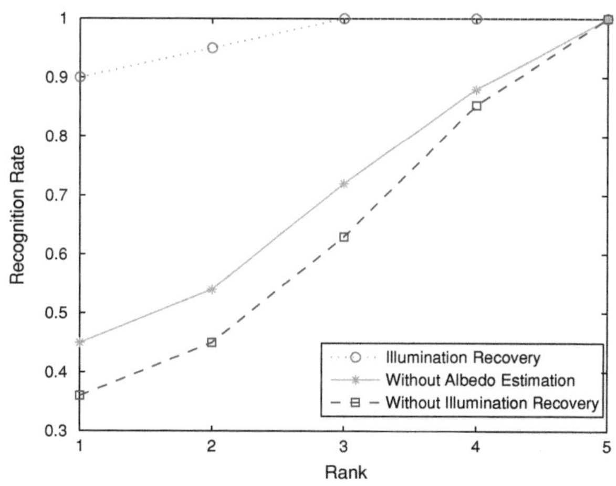

body as albedo (no albedo estimation), and (3) our framework. Figure 15 shows the recognition results, where the x-coordinate denotes the size of each gallery from 1 to 5, and the y-coordinate denotes the recognition rates. From the results, we can see that our framework always achieves the highest recognition rates among these methods. Besides, the performance of our framework is robust to the size of each gallery while the method without illumination recovery does not perform well when the size of the gallery is very small. This is because they do not consider the illumination variations, and thus it needs many more images to discern illumination changes. Furthermore, their method is also more restricted under some extreme poses, for example, an image taken from the front of a vehicle can never provide texture information on two sides and the back. However, our framework tremendously improves the recognition performance in these aspects, by which we can still recognize vehicles under limited inputs with unconstrained pose and illumination conditions.

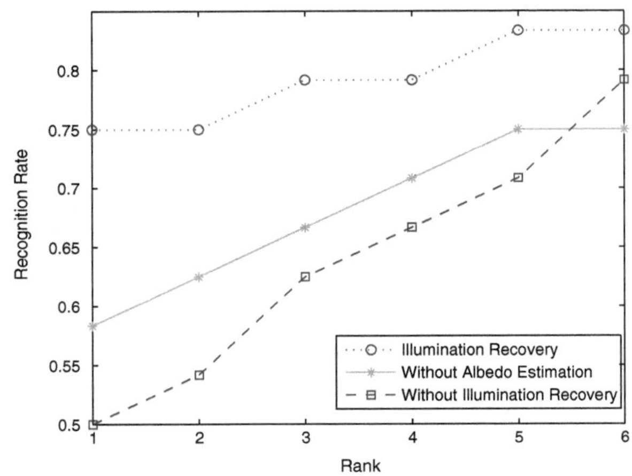

Fig. 16 Recognition result on real data (visible spectrum)

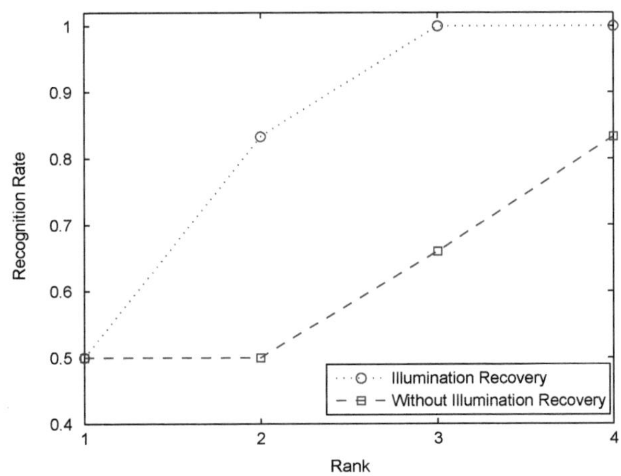

Fig. 17 Recognition result on real data (near-IR spectrum)

5.2.2 Real Data

Our real data consists of 30 image galleries captured from 30 vehicles (24 under visible illumination, and six under near-IR illumination). For visible illumination, each gallery has seven images under various viewpoints and lightings, while for IR illumination, there are five images in each gallery. All six galleries in near-IR illumination are from the same category of vehicle, since we have a small-size dataset. The image resolution is from 310×233 to 640×480. Our experiment is conducted by the same scheme as we did on synthetic data, and we do not mix up color images and near-IR images since they apparently different.

Figure 16 shows the recognition result of real data under visible illumination by the following methods: the method without illumination recovery, the method using average texture as albedo, and our framework with illumination recovery under different ranks. Figure 17 shows the near-IR part. The x-coordinate denotes the rank (number of images involved per gallery), and the y-coordinate denotes the recognition rates. From results, we can see that by illumination recovery, the recognition results of our methods are significantly improved and stable when the number of images involved per gallery changes. However, the other two methods use semantic ownership of vehicle model and the symmetry of vehicle body to represent texture information. These are not accurate due to the effect of illumination conditions, especially when the size of the gallery is small.

6 Conclusion

We have detailed a 3D model-driven framework to match vehicles subject to large variations of both pose and lighting in visible or near-IR illumination. By estimated pose and albedo, the illumination condition can be approximately recovered by using spherical harmonics representation. This will also allow us to re-light the reference object under any target condition of pose and illumination. Based on algorithmic components, matching between two input images is conducted in a common domain by computing the distance from the re-rendered images. Experimental results demonstrate that our framework has improved the matching and recognition performance, especially when objects are under both large pose and illumination variations. Besides vehicles, our framework can also be generalized to handle other types of objects.

There are also certain limitations in our framework. When the approximated fitting is coarse and inaccurate due to non-standard types of vehicles and camera distortion, the recognition suffers and is limited to vehicles subject to the same category. Besides, the real illumination condition is much more complicated than the current assumption, i.e., a dominating light source in infinity. For example, in an indoor environment such as auto-show sometimes the highlight area can affect the recognition results. The first limitation can be further improved by the morphable model, which has been successfully applied in face recognition; while the second limitation requires more techniques in the illumination model, which are the future tasks we expect to undertake.

Acknowledgments Hong Qin and Tingbo Hou (Stony Brook University)'s research reported in this chapter is partially supported by NSF Grants: IIS-0949467, IIS-0710819, and IIS-0830183.

References

1. Basri, R., Jacobs, D.W.: Lambertian reflectance and linear subspaces. TPAMI **25**(2), 218–233 (2003)
2. Biswas, S., Agrawal, G., Chellappa, R.: Robust estimation of albedo for illumination-invariant matching and shape recovery. In: ICCV (2007).

3. Debevec, P., Hawkins, T., Tchou, C., Duiker, H.-P., Sarokin, W., Sagar, M.: Acquiring the reflectance field of a human face. In: SIGGRAPH, pp. 145–156 (2000)
4. Fergus, R., Perona, P., Zisserman, A.: Object class recognition by unsupervised scale-invariant learning. In: CVPR, pp. 264–271 (2006)
5. Fredembach, C., Süsstrunk, S.: Colouring the near infrared. In: Proceedings of the IS & T/SID 16th Color Imaging Conference, pp. 176–182 (2008)
6. Fredembach, C., Süsstrunk, S.: Illuminat estimation and detection using near infrared. In: SPIE/IS & T Electronic Imaging, vol. 7250 (2009)
7. Gardner, W.F., Lawton, D.T.: Interactive model-based vehicle tracking. TPAMI **18**(11), 1115–1121 (1996)
8. Guo, Y., Rao, C., Samarasekera, S., Kim, J., Kumar, R., Sawhney, H.: Matching vehicles under large pose transformations using approximate 3d models and piecewise mrf model. In: CVPR (2008)
9. Guo, Y., Shan, Y., Sawhney, H., Kumar, R.: Peet: prototype embedding and embedding transition for matching vehicles over disparate viewpoints. In: CVPR (2007)
10. Hou, T., Wang, S., Qin, H.: Vehicle matching and recognition under large variations of pose and illumination. In: CVPR Workshop on Object Tracking and Classification Beyond and in the Visible Spectrum, pp. 24–29 (2009)
11. Ji, Q., Yang, X.: Real time 3d face pose discrimination based on active ir illumination. In: ICPR, vol. 4, pp. 310–313 (2002)
12. Kim, Z., Malik, J.: Fast vehicle detection with probabilistic feature grouping and its application to vehicle tracking. In: ICCV, pp. 524–531 (2003)
13. Koller, D., Daniilidis, K., Nagel, H.-H.: Model-based object tracking in monocular image sequences of road traffic scenes. IJCV **10**(3), 257–281 (1993)
14. Li, S.Z., Chu, R., Liao, S., Zhang, L.: Illumination invariant face recognition using near-infrared images. TPAMI **29**(4), 627–639 (2007)
15. Morris, N.J.W., Avidan, S., Matusik, W., Pfister, H.: Statistics of infrared images. In: CVPR (2005)
16. Murase, H., Nayar, S.K.: Visual learning and recognition of 3-d objects from appearance. IJCV **14**(1), 5–24 (1995)
17. Novotny, P.M., Ferrier, N.J.: Using infrared sensors and the phong illumination model to measure distances. In: ICRA, pp. 1644–1649 (1999)
18. Pan, K., Liao, S., Zhang, Z., Li, S.Z., Zhang, P.: Part-based face recognition using near infrared images. In: CVPR (2007)
19. Ramamoorthi, R., Hanrahan, P.: A signal-processing framework for inverse rendering. In: SIGGRAPH, pp. 117–128 (2001)
20. Romdhani, S., Blanz, V., Vetter, T.: Face identification by fitting a 3d morphable model using linear shape and texture error functions. In: ECCV, pp. 3–19 (2002)
21. Sato, Y., Wheeler, M.D., Ikeuchi, K.: Object shape and reflectance modeling from observation. In: SIGGRAPH, pp. 379–388 (1997)
22. Shan, Y., Sawhney, H.S., Kumar, R.: Vehicle identification between non-overlapping cameras without direct feature matching. In: ICCV, pp. 378–385 (2005)
23. Shan, Y., Sawhney, H.S., Kumar, R.: Unsupervised learning of discriminative edge measures for vehicle matching between non-overlapping cameras. TPAMI **30**(4):700–711 (2008)
24. Wang, O., Davis, J., Chuang, E., Rickard, I., de Mesa, K., Dave, C.: Video relighting using infrared illumination. In: Computer Graphics Forum (Proceedings Eurographics), vol. **27**(2) (2008)
25. Wang, Y., Liu, Z.C., Hua, G., Wen, Z., Zhang, Z.Y., Samaras, D.: Face re-lighting from a single image under harsh lighting conditions. In: CVPR (2007)
26. Wen, Z., Liu, Z., Huang, T.S.: Face relighting with radiance environment maps. In: CVPR, pp. 158–165 (2003)
27. Yu, Y., Debevec, P., Malik, J., Hawkins, T.: Inverse global illumination: recovering reflectance models of real scenes from photographs. In: SIGGRAPH, pp. 215–224 (1999)

28. Zhang, L., Samars, D.: Face recognition from a single training image under arbitrary unknown lighting using spherical harmonics. TPAMI **28**(3), 351–363 (2006)
29. Zhang, L., Wang, S., Samaras, D.: Face synthesis and recognition from a single image under arbitrary unknown lighting using a spherical harmonic basis morphable model. In: CVPR, pp. 209–216 (2005)
30. Zhao, S., Grigat, R.-R.: Robust eye detection under active infrared illumination. In: ICPR, vol. 4, pp. 481–484 (2006)
31. Zhu, Z., Ji, Q., Fujimura, K., Lee, K.: Combining Kalman filtering and mean shift for real time eye tracking under active ir illumination. In: ICPR, vol. 4, pp. 318–321 (2002)
32. Zou, X., Kittler, J., Messer, K.: Face recognition using active near-ir illumination. In: BMVC (2005)

Pattern Recognition and Tracking in Forward Looking Infrared Imagery

Mohammad S. Alam

Abstract In this chapter, we review the recent trends and advancements on pattern recognition and tracking in forward looking infrared (FLIR) imagery. In particular, we discuss several target detection and tracking algorithms for single/multiple target detection and tracking purposes. Each detection and tracking algorithm utilizes various properties of targets and image frames of a given sequence. At first we discuss a Fukunga–Kuntz Transform and template matching based algorithm for target detection and tracking. Then, we described a novel algorithm for target detection and tracking using fringe-adjusted joint transform correlation (JTC) and template matching. Finally, we discussed an invariant detection and tracking algorithm using a combination of fringe-adjusted JTC and a composited weighted reference function. The impact of sensor ego motion and possible compensation techniques as well as the role of image segmentation towards enhancing the accuracy of target detection and tracking is also described. The aforementioned techniques can detect and track small objects comprising of only a few pixels and is capable of compensating the high ego-motion of the sensor for various challenging scenarios. Test results obtained using real life FLIR image sequences are included to verify the effectiveness of the above mentioned algorithms for target detection and tracking in FLIR imagery.

Keywords Pattern recognition · Forward-looking infrared imagery · Target detection · Target tracking · Fringe-adjusted JTC · Correlation discrimination · Synthetic discriminant function · Invariant pattern recognition · Global motion compensation · Subframe segmentation

M. S. Alam (✉)
Department of Electrical and Computer Engineering, University of South Alabama,
Mobile, AL 36688-0002, USA
e-mail: malam@usouthal.edu

R. Hammoud et al. (eds.), *Machine Vision Beyond Visible Spectrum*,
Augmented Vision and Reality, 1, DOI: 10.1007/978-3-642-11568-4_4,

1 Introduction

One of the goals of automatic target recognition is to detect targets at long range. Long-range images are generally captured within the visible band of the electromagnetic wave spectrum because the sun emits all frequencies of radiation with peak around the visible band and radiation in the visible band can travel long distances in our atmospheric environment. However, visible band signals drop below the level of reliable detection when the sun is not shining overhead. Alternatively, long range images can also be recorded in the infrared (IR) region provided the target material spontaneously emits considerable amount of radiation energy in the infrared band. The images obtained by sensing the radiation in the infrared spectrum are also known as thermal images.

Although a great deal of effort has been expended on detecting objects in visual images, only limited amount of work has been reported on the detection and tracking of targets in infrared images. In general, existing methods on IR images work for a limited number of situations due to the constraints imposed on the solution and engineers tend to emulate human vision when designing imaging systems. In contrast to visual images, thermal images such as forward-looking infrared (FLIR) imagery usually have low resolution, low signal-to-noise ratio, low contrast, and very small target signature (a few pixels). Moreover, the variation of the target signature, competing background clutter, lack of a priori information, high ego-motion of the sensor, and artifacts generated due to the weather conditions complicate the detection and tracking of targets [1–10]. Almost all of the detection and tracking algorithms used for FLIR imagery assume that the target is brighter than the background and the signal-to-noise ratio and signal-to-clutter ratio are at acceptable levels. For example, a spatial filter based on the least mean square to maximize the signal-to-clutter ratio for a known clutter environment was introduced in Ref. [2], and a detector with constant false alarm rate using generalized maximum likelihood ratio was developed in Ref. [3]. An intensity variation function based technique for target detection in FLIR imagery was proposed in Ref. [4]. Another recent technique utilized fuzzy clustering, edge fusion and local texture energy for detecting targets in FLIR imagery [5]. Loo and Alam [6] used an invariant weighted composite reference function (WCRF) to represent target model in fringe-adjusted joint transform correlation based FLIR target detection and tracking. Bharadwaj et al. [7] classified FLIR image targets using hidden Markov trees. Cooper et al. [8] used a deformable template representation accommodating both geometric and signature variability for target detection in FLIR imagery. Recently, another technique for multiple target tracking via global motion compensation and optoelectronic correlation was proposed [9]. A decision fusion based algorithm for target tracking in FLIR image sequences was developed where the strategy is to prevent the development of various failure modes during tracking [10].

In this chapter, we discuss several target detection and tracking algorithms which are based on the Fukunaga–Koontz Transform (FKT), intensity variation function (IVF), normalized cross-correlation, template matching (TM), and fringe-adjusted joint transform correlation (fringe-adjusted JTC) and CRF based techniques. The performance of these algorithms was tested using 50 real life FLIR image sequences supplied by the Army Missile Command.

2 FKT Based Pattern Recognition and Tracking

Over the last few decades, object detection and tracking systems had been widely investigated using both digital and optical techniques. Digital techniques primarily involve statistical and artificial intelligence approaches [5, 11–16], while optical or optoelectronic techniques are primarily based on optical pattern recognition algorithms such as matched filtering and joint transform correlation [17–19]. Among the statistical target detection applications, recently, Yilmaz et al. [5] applied sensor ego motion compensation and kernel density estimation using probability density functions (*pdf*) of intensity values and local standard deviations. Strehl and Aggarwal [14] proposed a segmentation and Bayesian estimation based detection and motion estimation approaches. Mahalanobis et al. [15, 16] defined and utilized quadratic correlation filters for target detection and discrimination. Fukunaga–Koontz transform has shown excellent promise for target detection problems due to inherent shift invariance property and no pixel-based feature extraction or preprocessing are required [16]. Fukunaga–Koontz transform also provides the best low-rank approximation for the quadratic discriminant analysis. This feature is especially important to realize a low-rank approximation for target recognition applications since the number of basis functions determines the complexity of the algorithm [20].

For target tracking, FLIR imagery poses different types of challenges such as low signal-to-noise ratio, background clutter, unpredictable camera motion, target occlusion, and illumination variation due to weather condition. In general, the detection of moving targets is realized from a stationary or moving platform. If the camera is stationary, the problem may be addressed by detecting the presence of motion in the image sequence. However, when the camera is mounted on a moving platform, motion detection alone is insufficient, because the camera motion produces an effect that appears to be the motion of the background. Such motion related problems must be compensated to improve the accuracy of detection and tracking results [5, 21–23]. Besides these factors, in some scenes, smoke generated by the target may cause additional motion on the images of the frame, which further complicates the detection and tracking process. In this section, we discuss a regional motion compensation algorithm to overcome the detrimental effects of the camera motion, smoke motion, as well as other factors on the detection and tracking performance. The regional motion compensation technique incorporates target location based regional segmentation and template matching techniques [9]. The regional segmentation approach alleviates the effects of the smoke motion and reduces the computation burden because it does not require the processing of the whole image. In addition, the template matching technique enhances

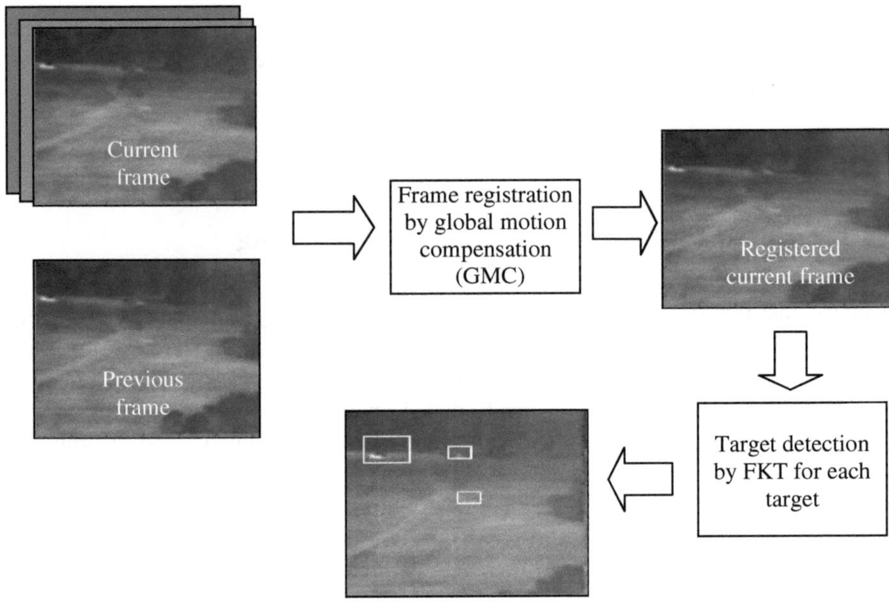

Fig. 1 The schematic diagram of the proposed multiple target tracking algorithm

processing speed and accurately estimates the target motion since it does not utilize any extra image preprocessing algorithms. Finally, using the estimation results, the image frame is recovered before applying the FKT based tracking algorithm.

In this section, we discuss a novel technique for both single/multiple target detection and tracking using a combination of regional motion compensation and FKT. The FKT provides the best eigenvectors to discriminate the target from background clutter. The basis functions are obtained separately for each class, i.e., targets and background, and their correlation outputs are independently processed to find the location of the target. To decrease the computation time and to avoid the undesired background clutter effects on the tracking processes, subframe segmentation is performed that has been widely utilized for previous target tracking algorithms [4]. The subframe segmentation of the current frame offers high speed processing capability while alleviating the detrimental effects of undesired background clutters. The FKT accurately indicates the location of the target(s) without requiring any extra feature extraction operations. The following subsections describe target detection and tracking algorithm involving regional motion compensation technique and FKT with subframe segmentation. The schematic diagram of the proposed target detection and tracking algorithm is shown in Fig. 1.

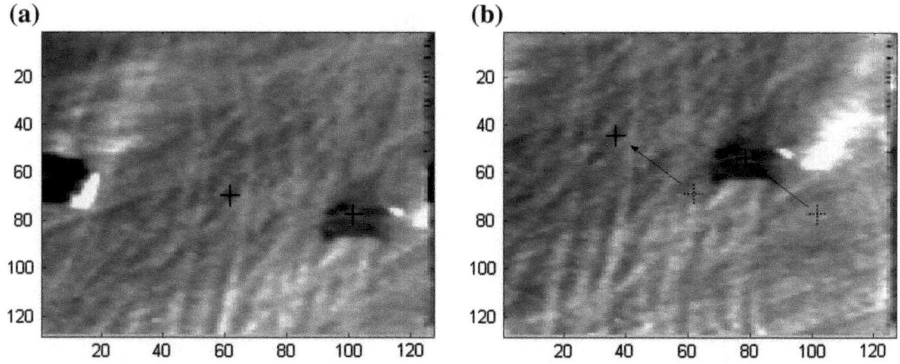

Fig. 2 The effect of global motion on the target location in a sample FLIE sequence (lwir_1816). **a** Target coordinates are $x = 101$ and $y = 77$ in the frame # 295, **b** target coordinates are $x = 78$ and $y = 53$ in the frame # 296

2.1 Global Motion Compensation

Camera motion estimation in image sequences generally focuses on the recovery of the frames when the camera is mounted on a moving platform. Global motion in video sequences is more complex and involves camera operation, camera motion and other non-target motions. Global motion compensation is usually handled by compensating the dominant motion using estimation and segmentation techniques. Ego motion related factors may significantly degrade the performance of any detection and tracking algorithm. Figure 2 demonstrates the global motion effects on the target region in a sample FLIR sequence where the location of target changes in consecutive frames (frames # 295–296 of sequence lwir_1816) due to camera motion. The 3D motion associated with an imaging system can be translated into 2D spatio-temporal changes of the gray level intensity function via regional segmentation. The regional segmentation is achieved by separating the image frame into equal sized regions. The reference image is selected from a database or on the basis of known target characteristics, which is then used as a reference in the template matching technique. As the motion estimation algorithm searches the region of interest, a subframe is selected around the above mentioned reference image which is 20-pixel larger than the reference image [4]. Figure 3 demonstrates the region selection and segmentation processes.

To estimate the shifting distance of pixels, the TM algorithm is employed which yields the highest value corresponding to the correlation coefficients between the known reference image and the unknown input scene. The correlation is a measure of association of the relationship between two images, defined as the ratio of the observed covariance of two normalized variables, divided by the highest possible covariance. When the observed covariance is as high as the possible covariance value, the correlation operation generates a value of 1, indicating a perfect match

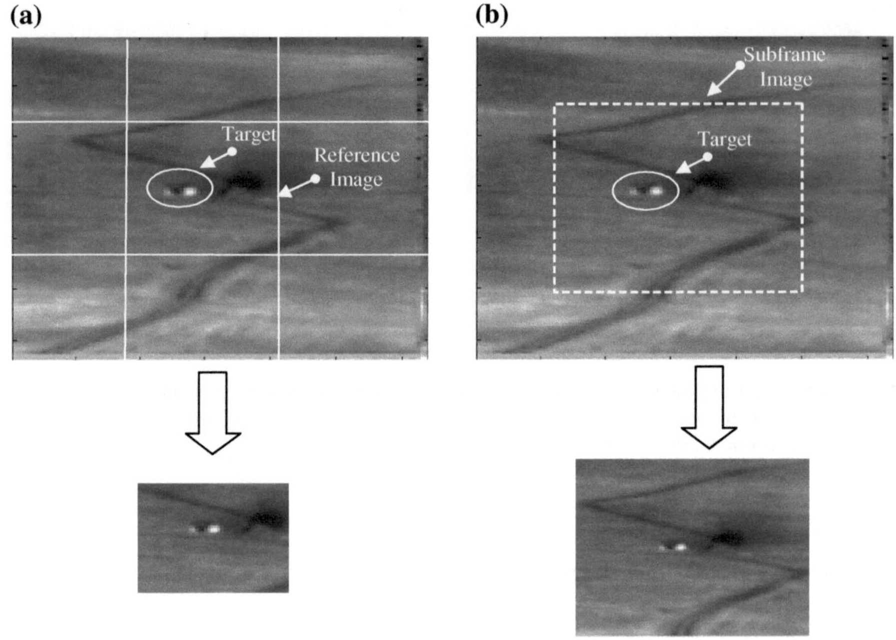

Fig. 3 Segmentation for regional global motion compensation in a longwave FLIR sequence (lwir_1815). **a** Reference image I_r segmentation from previous frame. **b** Subframe image I_s segmentation from current frame

between the two images. In contrast, a correlation value of 0 indicates a mismatch or random relationship between the two images [15].

Motion estimation may be used to determine the highest correlation between the reference image I_r obtained from the previous frame, and subframe image I_s obtained from the current frame as depicted in Fig. 3. The covariance of these two images may be defined as

$$c_{sr} = \sum_m \sum_n (I_s(m,n) - \mu_s)(I_r(m,n) - \mu_r) \tag{1}$$

where μ_s and μ_r are subframe image mean and reference image mean, respectively. The mean value for each image is defined as

$$\mu = \frac{1}{MN} \sum_1^N \sum_1^M I(m,n). \tag{2}$$

The comparison of the covariance variables is not easy because the variables may differ in magnitude (mean value) and dispersion (standard deviation). Therefore, normalized values are obtained with respect to the standard deviation of each variable, defined as

$$s_{sr} = \sqrt{\left(\sum_m \sum_n (I_s(m,n) - \mu_s)^2\right)} \times \sqrt{\left(\sum_m \sum_n (I_r(m,n) - \mu_r)^2\right)}. \tag{3}$$

Using Eqs. 1 and 3, the normalized correlation coefficient can be formulated as

$$r = \frac{c_{sr}}{s_{sr}}. \tag{4}$$

When the observed covariance represents the highest possible covariance, the correlation operation generates the maximum value, indicating perfect match between the reference image and the template image. Then, the shift distance of the reference image is calculated by using the following equations:

$$\begin{aligned} \Delta x &= x - \hat{x} \\ \Delta y &= y - \hat{y} \end{aligned} \tag{5}$$

where (x, y) represents the reference image coordinates and $(\hat{x}, \hat{y})$ denotes the coordinates of the maximum correlation value, defined as

$$(\hat{x}, \hat{y}) \in \max(r). \tag{6}$$

Thus the registration process is realized by shifting the entire subframe pixel coordinates by $(\Delta x, \Delta y)$ along the x- and y-axes, respectively [14].

2.2 Fukunaga–Koontz Transform

An image can be represented into two different classes, target and background clutter, set as normalized image vectors x and y, where the mean is subtracted from each image. Then, the covariance matrices can be calculated for the target and background classes as

$$\Gamma_1 = E[xx'] \tag{7}$$

and

$$\Gamma_2 = E[yy'] \tag{8}$$

where E denotes the expectation operation and x' and y' represents the transpose of each class image. The sum of these matrices can be represented as

$$\Gamma_1 + \Gamma_2 = \Phi\Lambda\Phi' \tag{9}$$

where Φ represents the eigenvector matrix and Λ represents the diagonal eigenvalue matrix. By combining these matrices, we can define a transformation operator T as

$$T = \Phi\Lambda^{-1/2} \tag{10}$$

Using the transformation operator T, the transformed new data set can be expressed as

$$\hat{x} = T'x \tag{11}$$

and

$$\hat{y} = T'y. \tag{12}$$

Then the sum of the covariance matrices for the transformed data becomes

$$T'(\Gamma_1 + \Gamma_2)T = I \tag{13}$$

where I represents the identity matrix. Thus the corresponding covariance matrices for each class can be calculated as

$$C_1 = T'\Gamma_1 T \tag{14}$$

and

$$C_2 = T'\Gamma_2 T \tag{15}$$

where C_1 denotes class one and C_2 represents class two. From Eqs. 14 and 15, it is obvious that

$$C_1 + C_2 = I \tag{16}$$

If $\vec{v}_i$ represents an eigenvector of C_1 corresponding to the eigenvalue λ_j, then

$$C_1\vec{v}_i = \lambda_j\vec{v}_i \tag{17}$$

Substituting Eq. 17 into Eq. 16, we obtain

$$C_2\vec{v}_i = (1 - \lambda_j)\vec{v}_i \tag{18}$$

From Eq. 18, it is obvious that C_2 has the same eigenvectors as C_1. However, the eigenvalues for C_1 is λ_j and the corresponding eigenvalues for C_2 is $(1-\lambda_j)$ implying that the two classes share complimentary eigenvalues as depicted in Fig. 4. This property of eigenvalues can be used for discriminating two classes for classification purposes.

The tuned basis functions (TBFs) are defined according to the complimentary distribution of eigenvalues. Initially, the eigenvalues are organized in ascending order for each class. Then the dominant eigenvectors corresponding to dominant eigenvalues for class C_1 is used to construct TBFs. If N_1 eigenvectors best represents class C_1, then the TBF vectors for C_1 may be obtained as

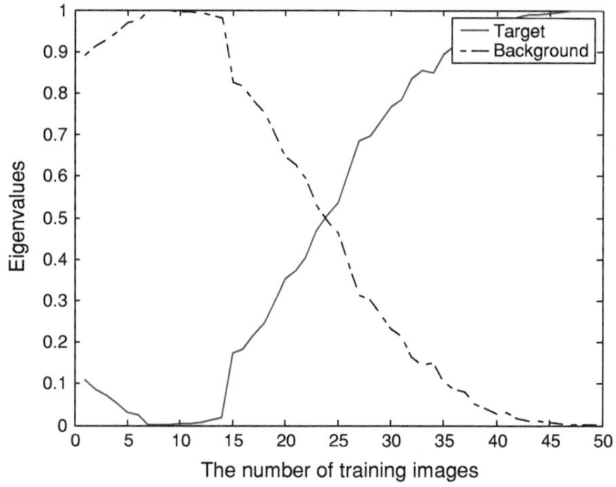

Fig. 4 Complimentary eigenvalues of two classes

$$V_1 = \left[\vec{v}_1 \; \vec{v}_2 \; \ldots \; \vec{v}_{N_1} \right] \tag{19}$$

Similarly, the rest of N_2 eigenvectors are used to construct TBFs for class C_2, defined as

$$V_2 = \left[\vec{v}_{N-N_2+1} \; \vec{v}_{N-N_2+2} \; \ldots \; \vec{v}_N \right] \tag{20}$$

From Eqs. 19 and 20, it is evident that $N_1 + N_2 = N$ [20]. For classification applications, reduced dimension of TBF vectors are preferred to decrease the computation burden. Therefore, the dominant eigenvectors are selected corresponding to the dominant eigenvalues for each class.

2.3 FKT Based Detection and Tracking

A Fukunaga–Koontz transform based multiple target detection algorithm is used to detect location of targets in each frame for tracking purposes. We also used subframe approach for high speed processing and improve tracking accuracy. The subframe segmentation step includes basic selection of region of interest as shown in Fig. 3b. In each frame, subframe location is updated corresponding to target coordinates obtained from the previous frame. In the target detection step, image chips selected by sliding in the subframe are converted to vector form. Each vector is then transformed by transformation operation T as

$$\hat{z}_{ijk} = T_k' z_{ij} \tag{21}$$

where ij represents image chip coordinates in the subframe, k denotes class number according to target classes, and the size of z_{ij} is equal to the size of x training images.

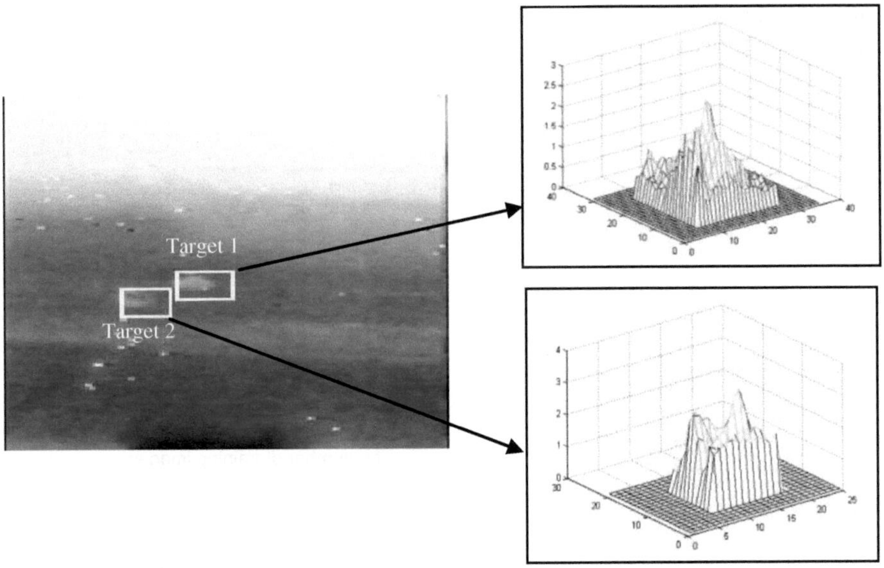

Fig. 5 Detection and tracking results for multiple targets in the same field of view

To formulate the detection metrics, the projection of $\hat{z}_{ijk}$ on the tuned basis functions can be computed as

$$\Omega_k = V_k' \hat{z}_{ijk} \tag{22}$$

and

$$\Omega_b = V_b' \hat{z}_{ijk} \tag{23}$$

where V_k represents TBFs for the kth target class, V_b represents TBFs of the background. In general, the TBFs of each target class incorporate more information about that target class. Therefore, the TBFs of background may be ignored for special cases to enhance processing speed. If the test image chip $\hat{z}_{ijk}$ corresponds to the possible target candidate, the detection metric Ω_k produces higher value compared to non-target object or background image chip values. The summation of each Ω_k is expected to be large for target(s) and small for background clutters as shown in Fig. 5. Thus, the coordinates of the highest value ij in the projection plane indicates the detected target coordinates in the subframe. According to the obtained projection plane h, target coordinates are determined as

$$(\hat{x}, \hat{y}) \in \max(h(x, y)). \tag{24}$$

The new coordinates $(\hat{x}, \hat{y})$ of the target are used to update the subframe location for the next frame. These steps are repeated until all frames of the sequence are processed.

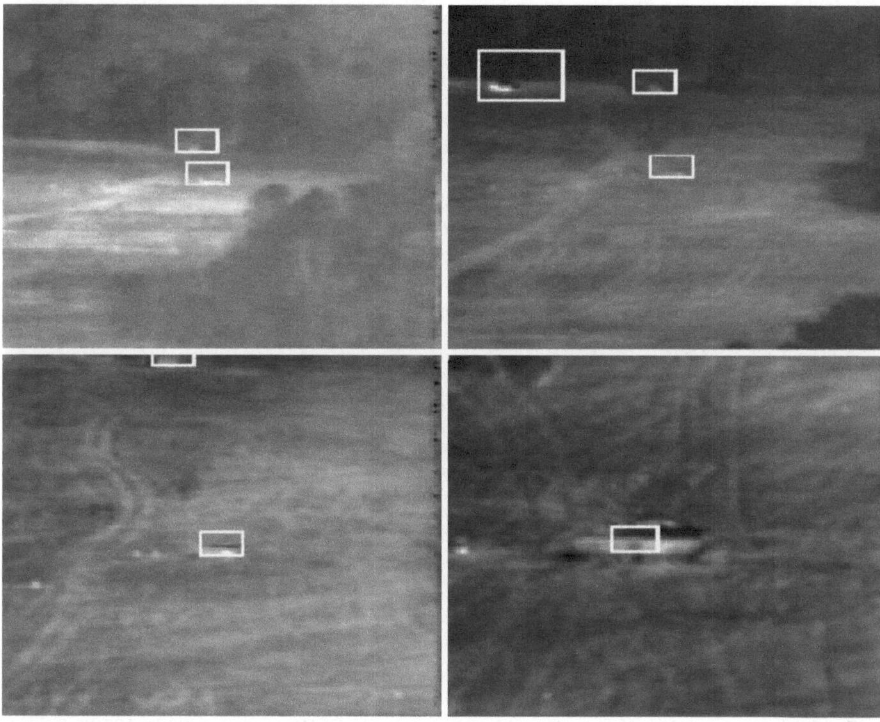

Fig. 6 Target detection and tracking results in a long wave FLIR sequence (lwir_1812) involving three targets with different challenging scenarios

2.4 Test Results

To test the performance of the proposed algorithm, real life FLIR image sequences supplied by the AMCOM are used. There are 50 FLIR sequences containing both single and multiple moving targets under various challenging scenarios such as the presence of target-like clutter as well as other artifacts. These image sequences were acquired by an infrared sensor installed on a moving aircraft. As the imaging system platform does not remain stable during the exposure time, all of the objects in the image sequence are impacted by the motion of the platform as well as other distortions. For testing purposes, we considered an arbitrary image sequence consisting of more than 300 frames where each frame consists of 128×128 pixels. A detailed software package was developed to test the performance of the FKT based algorithm.

To verify the performance of the proposed algorithm for single/multiple target detection and tracking, FLIR sequences were chosen with painstaking care for testing purposes. Figure 6 shows the results obtained for a longwave FLIR sequence (lwir_1812) involving three targets with different challenging scenarios. Initially, there are two targets (Bradley and tank) with low SNR and low contrast. At frame

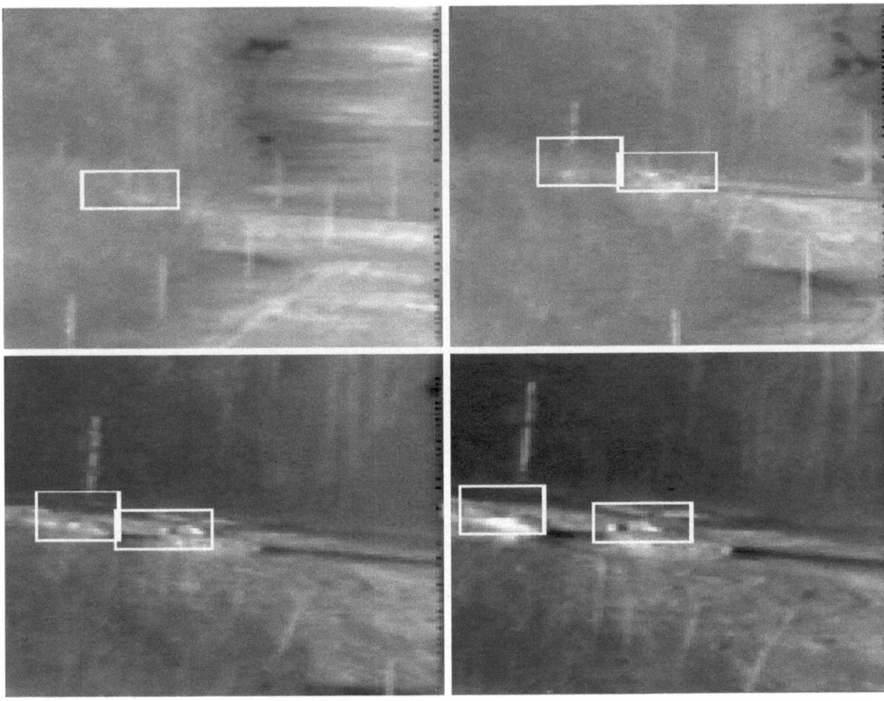

Fig. 7 Multiple target detection and tracking in a sample FLIR sequence (lwir_1910). Sample frames (7, 70, 102 and 130) showing the detection and tracking of two similar low contrast targets located in close proximity

#66, the third target (M60) enters into the scene which is a moving target in the sequence. Then the second and third targets move out of the field-of-view sequentially in the subsequent frames. It may be mentioned that near the end of the sequence, the first target changes scale due to camera zooming operation. However, the FKT-based algorithm accurately detects and tracks these three targets in all of the 300 frames without any ambiguity.

In the second example, we selected another challenging longwave FLIR sequence (lwir_1910) as shown in Fig. 7. Initially, the field of view contains only one low contrast target blended with the background as shown in frame #17 of the sequence. After frame #56, the second similar target appears in the field of view, which is located in close proximity with the first target. From Fig. 7, it is evident that the FKT-based detection and tracking algorithm successfully detects and tracks targets in FLIR sequences.

2.5 Conclusion

In this chapter, we described a new algorithm for multiple target tracking based on Fukunaga-Koontz transform. This algorithm enhances tracking performance by utilizing a regional global motion compensation technique. Since FLIR images are recorded from a moving platform, almost all of the frames are affected by camera motion and variability of the surroundings of targets, which includes illumination variation, shadowing, noise and occlusion. To obtain a robust tracking system, the ego motions are estimated using regional segmentation and template matching based regional global motion compensation technique.

3 Pattern Recognition via Optoelectronic Correlation

Frame recovery operation by estimating undesired motion can play a crucial role toward enhancing accuracy and overall performance of the detection and tracking system. In this section, a normalized correlation based regional template matching algorithm is used to accurately recover frames before the application of the detection and tracking algorithm. Then, a robust target detection tracking system is employed using a combination of fringe-adjusted JTC and template matching techniques. The joint transform correlation technique is used to detect a target optoelectronically, while template matching technique is performed digitally. To increase the detection and tracking system speed and alleviate the effects of clutter, a subframe segmentation and local deviation based image preprocessing algorithm is used. The performance of the single/multiple target detection and tracking system is tested using the real life FLIR video sequences discussed in the previous section.

In general, the detection of moving targets is realized from a stationary or moving platform. If the camera is stationary, the problem may be addressed by detecting the presence of motion in the image sequence. However, when the camera is mounted on a moving platform, motion detection alone may not be enough, because the camera motion produces an effect that appears to be the motion of the background. Such motion related problems need to be compensated in order to improve the accuracy of detection and tracking [5, 21–25]. Besides the above mentioned factors, in some scenes smoke generated by the target causes additional motion on the objects of the image frame, which further complicates the detection and tracking process. To overcome the detrimental effects of the camera motion, smoke motion as well as other factors on tracking performance, the global motion compensation algorithm discussed in Sect. 2.2 is used which utilizes regional segmentation and template matching techniques. The regional segmentation approach alleviates the effects of the smoke motion and reduces the computation time because it does not require the processing of the entire image. In addition, the template matching technique enhances processing speed and accurately estimates the target motion since it does not require any preprocessing operation. Finally, using estimation results, the frame image is recovered before applying the detection and tracking algorithm.

Optoelectronic pattern recognition may involve either a matched filter based correlator [26] or a joint transform correlator [27–31]. The joint transform correlation (JTC) technique has shown remarkable promise for real time matching and tracking applications [27–31]. Unlike matched filtering techniques, a JTC does not require any complex-valued filter fabrication while relaxing the meticulous positioning of all

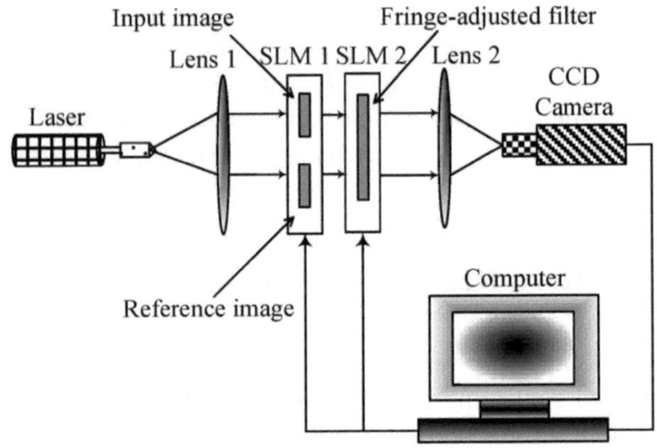

Fig. 8 Optoelectronic architecture for fringe-adjusted JTC based target tracking

components along the optical axis. Among the various JTC techniques, the fringe-adjusted JTC has been found to yield better correlation output [29–31]. Accordingly, in this section, we discuss a single/multiple target detection and tracking algorithm using a combination of fringe-adjusted JTC and template matching techniques.

The typical procedure utilized for detection and tracking involves a four-step process: subframe segmentation, feature extraction, classification, and tracking. The segmentation process is useful for dividing the image space into several smaller regions of interest. The feature extraction process allows the tracking system to identify and classify targets based on relevant features. Classification involves detecting and identifying the target(s) in question. Once the targets in an input scene are identified, they are tracked by correlating successive image frames using the fringe-adjusted JTC technique which yields excellent correlation peak corresponding to the known reference target and unknown input scene target.

Although the fringe-adjusted JTC technique produces robust results, occasionally it may fail due to background clutter effects or close similarity between the target and non-target objects. To ensure that the tracking system avoids wrong target detection, we utilized a normalized correlation based template matching (TM) algorithm [32]. The TM technique is initiated after evaluating the fringe-adjusted JTC output in a control module. The control module uses a Euclidian distance metric between the reference image and the new target candidate. If the control module triggers the TM algorithm, it searches for the maximal similarity between the reference target and the subframe using normalized cross-correlation coefficients. Thereafter, the target model is updated and is used as a reference for the subsequent frame [4]. This procedure is repeated until all frames of the sequence are processed.

3.1 Fringe-Adjusted JTC

The fringe-adjusted JTC technique has been found to yield better correlation output compared to alternate techniques for pattern recognition and tracking applications

[30, 31]. Unlike matched filtering, the JTC avoids complex filter fabrication, relaxes the meticulous alignment of system components, and provides near real-time parallel Fourier transformation of the reference image and the unknown input scene [32]. In a JTC, the reference image and the unknown input scene are introduced in the input plane by using a spatial light modulator (SLM) such as a liquid crystal television as shown in Fig. 8. The input plane SLM is illuminated by a parallel light beam generated by a coherent light source (laser) and a collimated lens (Lens L1). The input joint image of the JTC includes the reference and unknown input image, which can be formulated as

$$f(x, y) = r(x, y + y_0) + \sum_{i=1}^{n} t_i(x - x_i, y - y_i) + n(x, y - y_0) \tag{25}$$

where $r(x, y + y_0)$ represents the reference image, and $t(x, y - y_0)$ represents the input scene containing n targets $t_1(x - x_1, y - y_1), t_2(x - x_2, y - y_2), \ldots, t_n (x - x_n, y - y_n)$ respectively, which is corrupted by noise $n(x, y - y_0)$. In the first step, all pixels of SLM2 are set to 1. Lens L1 performs the Fourier transform of $f(x, y)$ and the intensity of the complex light distribution produced in the back focal plane of Lens 1, referred to as the joint power spectrum (JPS), is typically detected by a square-law detector such as a CCD array, given by

$$\begin{aligned}
|F(u, v)|^2 &= |R(u, v)|^2 + \sum_{i=1}^{n} |T_i(u, v)|^2 + |N(u, v)|^2 \\
&\quad + 2\sum_{i=1}^{n} |T_i(u, v)\|R(u, v)| \cos[\phi_{ti}(u, v) - \phi_r(u, v) - ux_i - vy_i - 2vy_0] \\
&\quad + 2\,|R(u, v)\|N(u, v)| \cos[\phi_r(u, v) - \phi_n(u, v) + 2vy_0)] \\
&\quad + 2\sum_{i=1}^{n} |T_i(u, v)\|N(u, v)| \cos[\phi_{ti}(u, v) - \phi_n(u, v) - ux_i - vy_i] \\
&\quad + 2\sum_{i=1}^{n} \sum_{\substack{k=1 \\ k \neq i}}^{n} |T_i(u, v)\|T_k(u, v)| \cos[\phi_{ti}(u, v) \\
&\quad - \phi_{tk}(u, v) - ux_i + ux_k - vy_i + vy_k]
\end{aligned} \tag{26}$$

where $|R(u, v)|$, $|T_i(u, v)|$, $|N(u, v)|$ are the amplitudes and $\phi_r(u, v)$, $\phi_{ti}(u, v)$, $\phi_n(u, v)$ are the phases of the Fourier transforms of $r(x, y)$, $t_i(x, y)$, $n(x, y)$, respectively; u and v are frequency-domain variables. In Eq. 26, the first three terms correspond to the zero order term, the fourth term correspond to the desired crosscorrelation between the reference image and the input scene targets, while the remaining terms correspond to the crosscorrelation between the reference image, noise and input scene targets, respectively. It may be mentioned that the presence of identical targets or non-target objects in the input scene yields undesired autocorrelation peaks or false alarms. To eliminate such false alarms, we used a Fourier plane image subtraction technique [30] where the input-scene-only power spectrum and the reference-image-only power spectrum are subtracted from the joint power spectrum of Eq. 26 before applying the inverse Fourier transform operation to yield the correlation output. After applying the Fourier plane image subtraction, the modified JPS may be expressed as

$$\begin{aligned} P(u,v) &= |F(u,v)|^2 - |T(u,v)|^2 - |R(u,v)|^2 \\ &= 2\sum_{i=1}^{n} |T_i(u,v)||R(u,v)| \cos[\phi_{ti}(u,v) - \phi_r(u,v) - ux_i - vy_i - 2vy_0] \\ &\quad + 2\,|R(u,v)||N(u,v)| \cos[\phi_r(u,v) - \phi_n(u,v) - 2vy_0] \end{aligned} \tag{27}$$

From Eq. 27, it is evident that the subtraction operation eliminates the false alarms generated by the similar input scene targets as well as the crosscorrelation terms between other objects that may be present in the input scene.

A classical JTC [33–36] usually yields large correlation side lobes, large correlation peak width, a strong zero-order peak and low optical efficiency. To alleviate the limitations associated with classical and binary JTCs, the fringe-adjusted JTC technique has been proposed where JPS is multiplied by a fringe-adjusted filter (FAF) before applying the inverse Fourier transform operation to yield the correlation output. The fringe adjusted filter is defined as

$$H(u,v) = C(u,v)\left[D(u,v) + |R(u,v)|^2\right]^{-1} \tag{28}$$

where $C(u,v)$ and $D(u,v)$ are either constants or functions of u and v. When $C(u,v) = 1$ and $|R(u,v)|^2 \gg D(u,v)$, the fringe-adjusted filter function $H(u,v)$ can be approximated as

$$H(u,v) \approx |R(u,v)|^{-2} \tag{29}$$

The modified JPS of Eq. 27 is then multiplied by the FAF of Eq. 29 to yield the modified fringe-adjusted JPS, given by

$$O(u,v) = H(u,v) \times P(u,v) \approx |R(u,v)|^{-2} \times P(u,v) \tag{30}$$

An inverse Fourier transform of Eq. 30 yields the correlation output that consists of a pair of crosscorrelation peaks corresponding to the known reference target and unknown input scene target(s) as shown in Fig. 9. The multiplication in Eq. 30 is achieved by displaying the JPS and the FAF in SLM1 and SLM2, respectively, and then illuminating these SLMs with the collimated light beam. The second lens (Lens L2) performs an inverse Fourier transform of the fringe-adjusted JPS to yield the correlation output which consists of a pair of delta-function-like correlation peaks corresponding to the input scene target located at $(\pm x_i, \pm y_i \pm 2y_0)$, and negligible cross-correlation output between the reference and non-target objects as well as the noise term. Then, the new target(s) coordinates are determined using the coordinate(s) of the highest peaks in the correlation output plane.

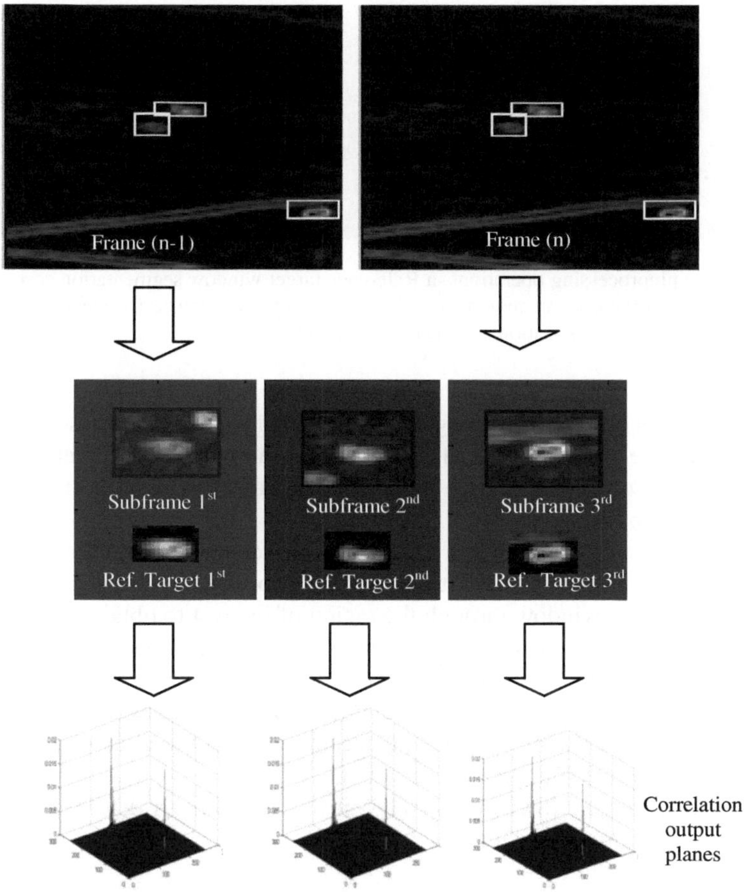

Fig. 9 Fringe-adjusted JTC target detection for three different targets

3.2 Target Detection and Tracking

The performance of any detection and tracking algorithm directly depends on the target signature, clutter, background, occlusion as well as other factors such as illumination. Image preprocessing plays an important role for enhancing the performance of the detection and tracking algorithm. Image segmentation is first step of preprocessing, where the whole image is divided into small regions of interest based on potential target locations. Working with a small region of interest significantly decreases the computation time while improving the robustness of the detection and tracking process [31, 37–45]. In the first frame of the sequence, the target is detected using a priori information from a database or based on known target characteristics. Based on the target information in the first frame, a reference target window is

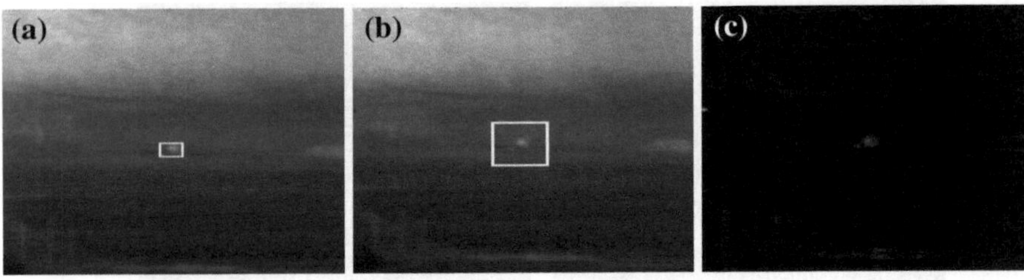

Fig. 10 Image preprocessing operations. **a** Reference target window segmentation corresponding to target size, **b** subframe segmentation which is 5 pixels larger than the target size, **c** result obtained after local deviation, normalization and mean subtraction

selected such that it is 2–3 pixels larger than the actual target size as shown in Fig. 10a. Subsequently, the subframe is gradually enlarged such that it is 5 pixels larger than the target window using reference target coordinates as the center of subframe as shown in Fig. 10b [4, 5].

After subframe segmentation, a local deviation technique is applied to suppress background clutter and to enhance target feature information. This technique mainly uses a small sliding window through the region of interest to obtain the differences between neighboring pixels. The local deviation technique generates nearly zero intensity pixels for smooth textures whereas high intensity pixels for edge regions which correspond to high frequency components. This improves the fringe-adjusted JTC based detection performance while accommodating the effects of other detrimental factors such as illumination variation.

The local deviation $L(x)$ of a pixel x in the subframe is determined from its neighborhood of M pixels, defined as

$$L(x) = \sqrt{\frac{1}{M-1}\sum_{x_i \in M} (I(x_i) - I(x))^2} \tag{31}$$

where I is the intensity value of pixels, x_i are the spatial location of the neighborhood pixels, and M represents the number of pixels in the neighborhood [42]. In this algorithm, we selected M as a 3×3 window. Finally, normalization and mean subtraction are applied to the reference and subframe images to assist in better visual interpretation and to simplify the target detection without losing relevant information. Normalization is realized by dividing each pixel value by the maximum pixel value in the subframe. Then, the mean value is subtracted from each pixel value of the subframe. From Fig. 10, it is obvious that the proposed preprocessing technique effectively enhances the distinctive features between the target and background while alleviating effects of the aforementioned detrimental factors.

A tracking technique may generate false alarms due to close similarity between the target and non-target objects. To ensure that the tracking system avoids wrong target detection, template matching technique, is used. The initiation of the TM algorithm depends on the output of a control module. The control module uses a Euclidian

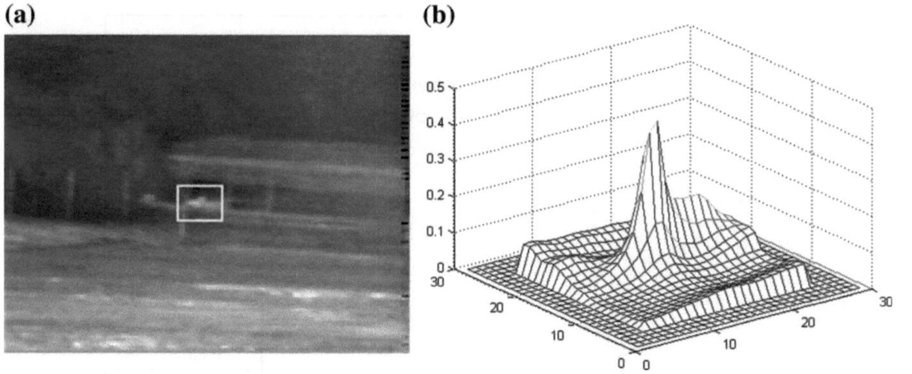

Fig. 11 **a** The first frame of a longwave sequence (L_1913), and **b** correlation output plane obtained using the TM algorithm

distance metric between the reference image and the new target candidate indicated by the fringe-adjusted JTC algorithm. The Euclidian distance metric is defined as

$$\rho_x = |x - \bar{x}|, \quad \rho_y = |y - \bar{y}|, \tag{32}$$

where (x, y) represents candidate target coordinates and $(\bar{x}, \bar{y})$ denotes reference target coordinates. When $\beta < \sqrt{\rho_x^2 + \rho_y^2}$, the TM technique is initiated, where β is a constant. The value of β is chosen by considering the computational complexity and the sensitivity needed for tracking the target.

The correlation coefficients between the reference and subframe images are calculated using Eq. 4. Then, the new target coordinates $(\hat{x}, \hat{y})$ in the subframe may be obtained by using Eq. 5, where $(\hat{x}, \hat{y})$ denote the coordinates corresponding to the maximum correlation peak in the TM output plane. Figure 11 shows the results obtained using the TM algorithm for one frame of an arbitrary FLIR sequence (L_1913). Since the target coordinates are obtained from the subframe, it is necessary to transform it to corresponding coordinates in the current frame. The target coordinates from the subframe location can be converted into the corresponding location in the current frame as

$$x^n = x^{n-1} + \hat{x} - \frac{\Gamma}{2} + 1 \tag{33}$$

and

$$y^n = y^{n-1} + \hat{y} - \frac{\Gamma}{2} + 1, \tag{34}$$

where (x^n, y^n) represent the new coordinates of the target in the nth frame (x^{n-1}, y^{n-1}) represent the coordinates of the target in the $(n-1)$th frame, Γ denotes the subframe size, and $(\hat{x}, \hat{y})$ denote the coordinates corresponding to the maximum correlation peak value in the correlation output plane [42]. Using the new coordinates

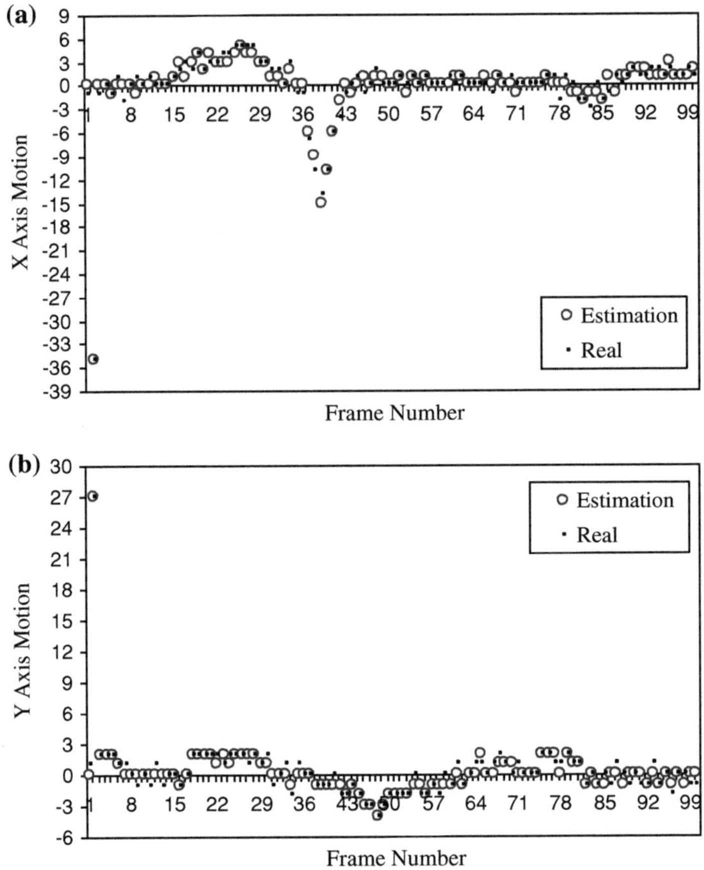

Fig. 12 Results obtained using the global motion compensation algorithm (denoted by *circle*) compared to the real coordinates of the frame (denoted by *dot*), **a** X-axis motion compensation results, and **b** Y-axis motion compensation results

of the target in the nth frame, the reference target image is updated. This reference target image is utilized to detect the new target coordinates in the subsequent $(n+1)$th frame.

3.3 *Test Results*

To evaluate the performance of the proposed algorithm, we used gray level real-world FLIR image sequences, supplied by AMCOM. As the imaging system platform is not stable during the exposure time, all of the objects in the image sequence are affected by the motion of the platform as well as other distortions. For testing purposes, we

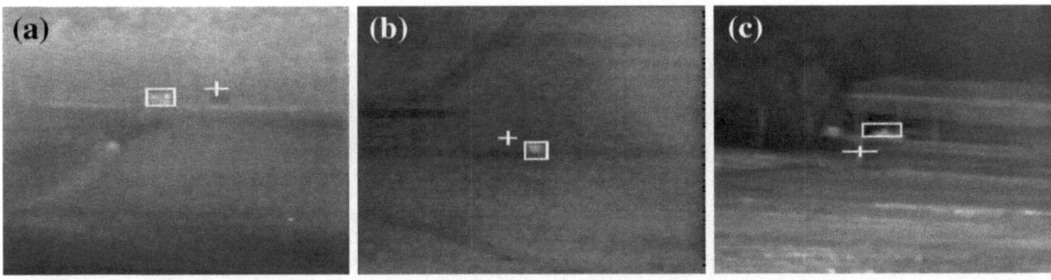

Fig. 13 The effect of global motion compensation on tracking performance, where "+" sign represents wrong tracking due to global motion and the box represent correct tracking after the global motion compensation; **a** Frame #2 of Sequence 1 (L_1701), **b** Frame #144 of Sequence 2 (L_2312), and **c** Frame #6 of Sequence 3 (L_1913)

considered an arbitrary image sequence consisting of more than 300 frames where each frame consists of 128 × 128 pixels.

In this algorithm, at first, each frame is recovered by using the global motion estimation algorithm. The results obtained using the global motion compensation algorithm compared to the actual location of the images are shown in Fig. 12. It is obvious from Fig. 12 that the proposed technique is effective in recovering the motion affected images. In some frames, the estimation algorithm generated 1 or 2 pixels of error margin, which can be neglected for target detection purposes.

Figure 13 demonstrates the effect of global motion compensation on the performance of the tracking algorithm for three arbitrary longwave FLIR sequences (L_1701, L_2312, and L_1913). In Fig. 13, "+" sign represents wrong tracking due to global motion related problems while the rectangular box represents correct tracking after the application of global motion compensation algorithm. It is evident that the application of the global motion compensation algorithm significantly enhances the detection and tracking performance.

After the global motion compensation, each image frame is preprocessed which primarily involves the segmentation operation to divide each frame into multiple subframes in order to detect the target(s) of interest. The local deviation and normalization process may be used to assist in better visual interpretation and to simplify subsequent target detection and tracking operations without losing relevant information. In the second step, fringe-adjusted JTC is employed to detect and to track each target. The fringe-adjusted JTC output is then fed to the control module which may trigger the TM technique depending on the results obtained from the fringe-adjusted JTC module. In this algorithm, the TM algorithm is utilized if and only if the fringe-adjusted JTC algorithm fails to detect and track a target. After detecting each target in a given frame, target reference windows and target coordinates are updated for subsequent frames, and the aforementioned steps are repeated.

To evaluate the performance of this algorithm for target detection and tracking, FLIR sequences were chosen with painstaking care. Figure 14 shows the results for three FLIR sequences involving multiple targets with different challenging scenarios such as scale and rotation variations, low SNR of the target, and high global motion

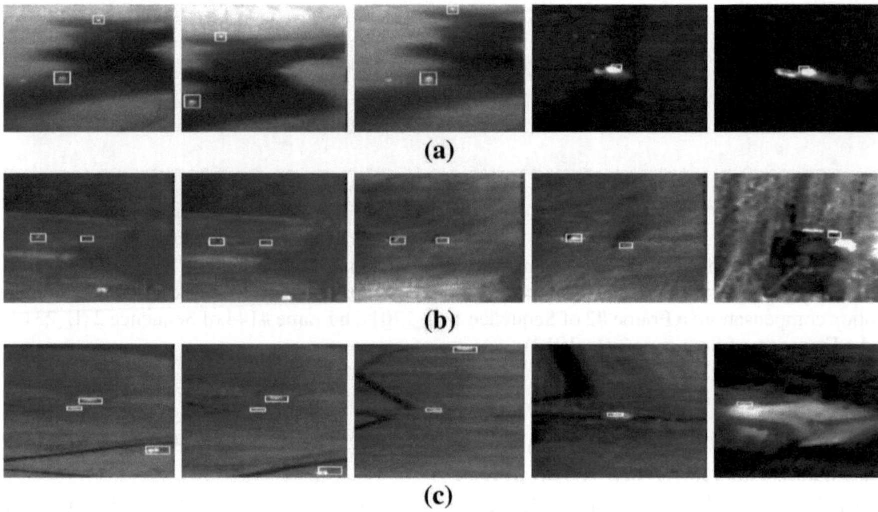

Fig. 14 Multiple target tracking results: **a** two moving targets affected by smoke movement, camera zooming, and camera motion in a longwave sequence (L_1720), frames 2, 12, 20, 680 and 778 are shown, **b** two moving targets in a longwave sequence (L_1816) affected by camera zooming, low contrast, and blending with the background, frames 1, 50, 150, 200, 328 are shown, **c** close proximity between two similar targets and camera zooming effect in a longwave sequence (L_1618), frames 1, 18, 100, 200 and 300 are shown

affected cases. In Fig. 14a, tracking results for two moving targets in a sequence (L_1720) are demonstrated. This sequence (L_1720) contains smoke movement, camera zooming and camera motion as well as other artifacts. It is obvious that this technique accurately tracks these two targets for all of the 778 frames of this sequence (L_1720). Figure 14b demonstrates the performance of the detection and tracking algorithm for tracking a low contrast target blended with the background in another challenging sequence (L_1816). Although the target size changes due to zooming of the camera, this algorithm successfully detects and tracks the two targets. Figure 14c demonstrates the results obtained for one more sequence (L_1618) involving similar targets which are located in close proximity. It is evident that the algorithm successfully detects and tracks all of the targets in the 300 frames of this sequence (L_1618).

The performance of the proposed algorithm has been tested for 50 different FLIR video sequences involving various distortions and challenging backgrounds/clutter. Table 1 shows the tracking result comparison between the fringe-adjusted JTC-TM algorithm discussed in this section and alternate techniques [5, 42]. From Table 1, it is evident that the proposed technique yields significantly better result when compared to the recently reported techniques. Table 2 illustrates the false alarm rate for the fringe-adjusted JTC technique and the fringe-adjusted JTC-TM algorithm.

Table 1 Tracking results comparison between the proposed algorithm and other techniques

Algorithms	Total number of sequence	Total number of targets	Tracking rate (%)
Algorithm of Ref. [5]	50	87	65
Algorithm of Ref. [42]	50	87	78
Fringe-adjusted JTC-TM	50	87	93

Table 2 False alarm ratio for the fringe-adjusted JTC technique and the fringe-adjusted JTC-TM algorithm

Sequence name	Total number of frames in each sequence	Number of false alarms by the fringe-adjusted JTC algorithm	False alarm ratio of the fringe-adjusted JTC algorithm (%)	Total number of frames tracked by the fringe-adjusted JTC-TM
L_1817S1	230	3	1.3	230
L_1701	388	7	1.8	388
L_2312	368	7	1.9	368
L_1913	265	6	2.2	265
L_1816	328	12	3.6	328
L_1415	281	14	4.9	281
L_1910	130	8	6.1	130
L_1720	778	73	9.3	778
L_2117	360	51	14.1	360
M_1407	400	71	17.7	400
L_1520	215	51	23.7	215

From Table 2, it is obvious that the fringe-adjusted JTC-TM algorithm effectively compensates the false alarms otherwise generated in a few frames by the fringe-adjusted JTC algorithm.

3.4 Conclusion

In this section, a fringe-adjusted JTC and TM based detection and tracking algorithm is discussed for single/multiple target in FLIR video sequences. This technique enhances detection and tracking performance by utilizing a global motion compensation technique. Since FLIR images are recorded from moving platform, almost all of the frames are affected by camera motion and variability of the surroundings of targets, which includes illumination variation, shadowing, noise and occlusion. In addition, camera zooming and smoke movement impacts the performance of the detection and tracking algorithm. To obtain a robust tracking system, the camera and smoke effects are eliminated using regional segmentation and template matching based global motion compensation technique. The regional segmentation of the compensation technique offers high speed processing capability while alleviating the detrimental effects of smoke movement scattered around the targets. The template matching technique accurately indicates the original location of the target without requiring any extra feature extraction operations.

In the second step, fringe-adjusted JTC and TM based optoelectronic detection and tracking system are utilized. The combination of these two techniques offers robust tracking performance in challenging scenarios. When the fringe-adjusted JTC algorithm yields consistently poor results due to blending or background clutter problems, template matching algorithm is employed as a compensation technique. From the test results obtained for various FLIR video sequences, it is evident that the fringe-adjusted JTC-TM algorithm yields excellent performance for near real time tracking of moving single/multiple target detection and tracking. The region based tracking approach instead of using the whole frame-based tracking techniques and the possibility of simultaneous multiple targets tracking using multiple processors enhances the processing speed for real applications. This algorithm can be implemented either digitally or by an optoelectronic implementation.

4 Invariant Object Detection and Tracking

To detect targets in FLIR image sequences, majority of the existing algorithms rely on the hot spot technique, which assume that the target radiation is much stronger than the radiation of the background and noise [3–5, 42–45]. However, in realistic detection and tracking scenarios, targets do not always appear brighter than the background. In contrast, motion detection methods based on motion tracking require background textures to differentiate the target movements [5, 14, 42]. Therefore, these techniques cannot effectively detect independently moving objects when the background contrast is low. In addition, motion detection methods yield false alarms caused by moving clutter and cannot effectively track non-moving targets. Similarly, some target tracking algorithms work only for sequences with no sensor ego-motion; where ego-motion means the transition of the entire captured image caused by the movement of the camera [25]. Other tracking methods [5, 43], which are based on mean-shift tracking rely on the intensity distributions generated from the target region and computes the translation of center of the target in the image space. These algorithms are more efficient as they do not confine to bright targets and are able to compensate for sensor ego-motion. In general, most of the aforementioned methods are computation intensive and are not suitable for real-time applications.

In this section, a novel approach is discussed for real-time target detection and tracking in FLIR imagery in the presence of high global motion as well as various distortions in target features. This technique utilizes fringe-adjusted JTC technique for real time estimation of target motion. For optical pattern recognition applications, the JTC technique has been found to be a robust alternative to VanderLugt type correlation. Although a classical JTC preserves the shift-invariant property, it suffers from high sensitivity to scale and rotation variations, large correlation side-lobes, and a high zero-order peak. Various improved versions of JTCs have been developed such as the binary JTC [33], amplitude modulated filter based JTC [46], and fringe-adjusted filter based JTC [47, 48]. Among these techniques [47–52], the fringe-adjusted JTC has been found to yield better correlation performance while alleviating the problems associated with alternate techniques.

In order to track a target in an input scene, a known reference image and the unknown input scene are introduced side-by-side in the input joint image for fringe-adjusted JTC processing. To detect and track a target in a FLIR image sequence, the reference image corresponding to the target to be detected is obtained from a database and the input image is obtained from the segmented initial frame and subsequent frames of the sequence. This algorithm is capable of tracking both moving and stationary targets while compensating the 3D distortions and ego-motion caused by the sensor. In addition, it can also track low-contrast objects and very small objects involving only a few pixels. The proposed technique has been tested with real FLIR imagery supplied by AMCOM and has been found to yield excellent performance.

Because FLIR imagery is generally corrupted by noise, a target may appear dissimilar from frame to frame. A frame is considered a bad frame by the tracking algorithm if it is corrupted by noise to the extent that the algorithm fails to track the target. Discarding the bad frame might solve the problem but the selection and determination of whether a particular frame is to be discarded requires the setting and application of a criterion or threshold. Selecting a wrong threshold value may cause the tracking algorithm to discard too many frames and, therefore frame discarding may not lead to a reliable solution. In this section, we utilized the concept of weighted composite reference function in the fringe-adjusted JTC based tracking algorithm that alleviates the effects of noise, 3D distortions, and hazards generated by bad frames yielding significantly improved detection and tracking performance.

4.1 Fringe-Adjusted JTC Based Motion Estimation

Target detection using the fringe-adjusted JTC technique has been discussed in Sect. 3.1. The location of the target in the unknown input scene is determined by examining the location of the maximum correlation peak in the output plane. For a moving target, target motion can be estimated by calculating the coordinates of the location of the target in the current frames and the subsequent frames.

4.2 Weighted Composite Reference Model

In general, FLIR images recorded using existing infrared cameras generate low-resolution noisy images, and target signatures may appear very different from one frame to another. It is necessary to identify the target regardless of the distortions, which could possibly occur and to reject non-target objects with low false alarm rate while ensuring a high probability of detection. To accommodate the problem of target signature variation due to in-plane/out-of-plane rotation and scale variations, partial occlusion, noise or bad frames, an enhanced version of the SDF [50] concept is used. Conventional SDF formulation uses a training set of images, which characterizes the expected geometric distortions. The training set images are used to reduce the filter's sensitivity to object distortions. An enhanced SDF formulation is employed in this chapter, called the weighted composite reference function, which preserves the information available from the preceding target reference images to reduce the effects of noise, hazard associated with bad frames, as well as other artifacts [10].

A weighted composite reference image can be generated by summing the multiplication of reference images of the target used in the previous frames with a set of weighted coefficients as shown in the following equation:

$$h_n(x, y) = a_0 s_0(x, y) + a_1 s_1(x, y) + a_2 s_2(x, y) + \cdots + a_{n-1} s_{n-1}(x, y), \quad (35)$$

where $s_i(x, y)$ is the ith reference image with $s_0(x, y)$ representing the first reference image, and $a_0, a_1, \ldots, a_n$ are a set of weighted coefficients. The composite SDF of Eq. 35 is designed in such a way that if all consecutive reference images are identical, i.e., $s = s_0 = s_1 = \cdots = s_n$, then the corresponding SDF composite image would be the reference image, i.e., $h_n = s$. Under the aforementioned condition, the summation of all coefficients becomes equal to 1. To eliminate the need for preserving all of the preceding reference images in the formulation of the WCRF shown in Eq. 35, a recursive WCRF model may be used, which is defined as

$$\begin{aligned} h_0(x, y) &= s_0(x, y) \\ h_i(x, y) &= \frac{1}{(k+1)} \left[k \times h_{i-1}(x, y) + s_i(x, y) \right], \quad i \leq n \end{aligned} \quad (36)$$

where k is an arbitrary constant. Rewriting Eq. 36 in the form of a series of summation yields

$$h_n(x, y) = \frac{k^{n-1}}{(k+1)^{n-1}} s_0(x, y) + \sum_{i=1}^{n-1} \frac{k^{n-i-1}}{(k+1)^{n-i}} s_i(x, y). \quad (37)$$

Comparing Eqs. 35 and 36, we get

$$\begin{aligned} a_i &= \frac{k^{n-i-1}}{(k+1)^{n-i}}, \quad 0 < i < n \\ &= \frac{k^{n-1}}{(k+1)^{n-1}}, \quad i = 0 \end{aligned} \quad (38)$$

From Eq. 38, it is evident that

$$\frac{k^{n-1}}{(k+1)^{n-1}} + \sum_{i=1}^{n-1} \frac{k^{n-i-1}}{(k+1)^{n-i}} = 1 \quad (39)$$

The larger the value of k, the more information from prior reference images is incorporated in the WCRF formulation. This makes the tracking algorithm less vulnerable to the effects of bad frames. In addition, it becomes less susceptible to the changes in target appearance with respect to various 3D distortions such as rotation and scale variations. For example, if k is chosen to be equal to 2, the SDF composite image generated at frame #5 of a given image sequence will be

$$h_5(x, y) = \frac{16}{81} s_0(x, y) + \frac{8}{81} s_1(x, y) + \frac{4}{27} s_2(x, y) + \frac{2}{9} s_3(x, y) + \frac{1}{3} s_4(x, y) \quad (40)$$

If the target at frame #4 is totally corrupted by noise and the tracking algorithm misses the target at frame #4, then according to the SDF reference image of Eq. 40, 33% of the target signal will be lost. With the remaining 67% of the target reference signal, we have 67% chance of tracking the right target at frame #5. Similarly, if the target in frame #3 is also corrupted by noise, we have 44% chance of tracking the right target at frame #5.

Figure 15 illustrates an example of WCRF assistance in the remedy of mistracking of a truck caused by a bad frame. Images in the left column are the input FLIR images where the tracked target is shown within a black box. Images in the middle column are the segmented reference images of the corresponding frame, while images in the right column correspond to the associated SDF composite images. In this FLIR sequence, frame #4 is totally corrupted by noise which results in the mistracking of the target. Conventional tracking methods [15–17], which utilizes only the target information from a single preceding frame, will have serious difficulty to track the target in subsequent frames without discarding the bad frame. However, the proposed tracking method using WCRF based target model update effectively tracks the target following the appearance of the bad frame.

4.3 Fringe-Adjusted JTC Based Target Tracking

Target tracking is more challenging than target detection because tracking involves the detection in consecutive frames while accommodating the effects of various 3D distortions, occlusion, noise as well as other artifacts. In general, a tracking algorithm identifies and tags the right target in a sequence of frames regardless of the target's signature variations. The proposed fringe-adjusted JTC based target tracking algorithm utilizes a priori information of the target from the previous frames such as the location of the target and the surrounding environment. The proposed tracking technique uses multiple epochs of the fringe-adjusted JTC tracking algorithm to zoom on and track the target while ensuring that the tracked target is the desired target.

In FLIR imagery, targets may consist of just a few pixels, which make the recognition of the target in the input scene challenging. Conventional way of solving this problem [38, 40] is to divide the input frame into several smaller subimages and apply the detection algorithm to individual segmented subimages in order to detect

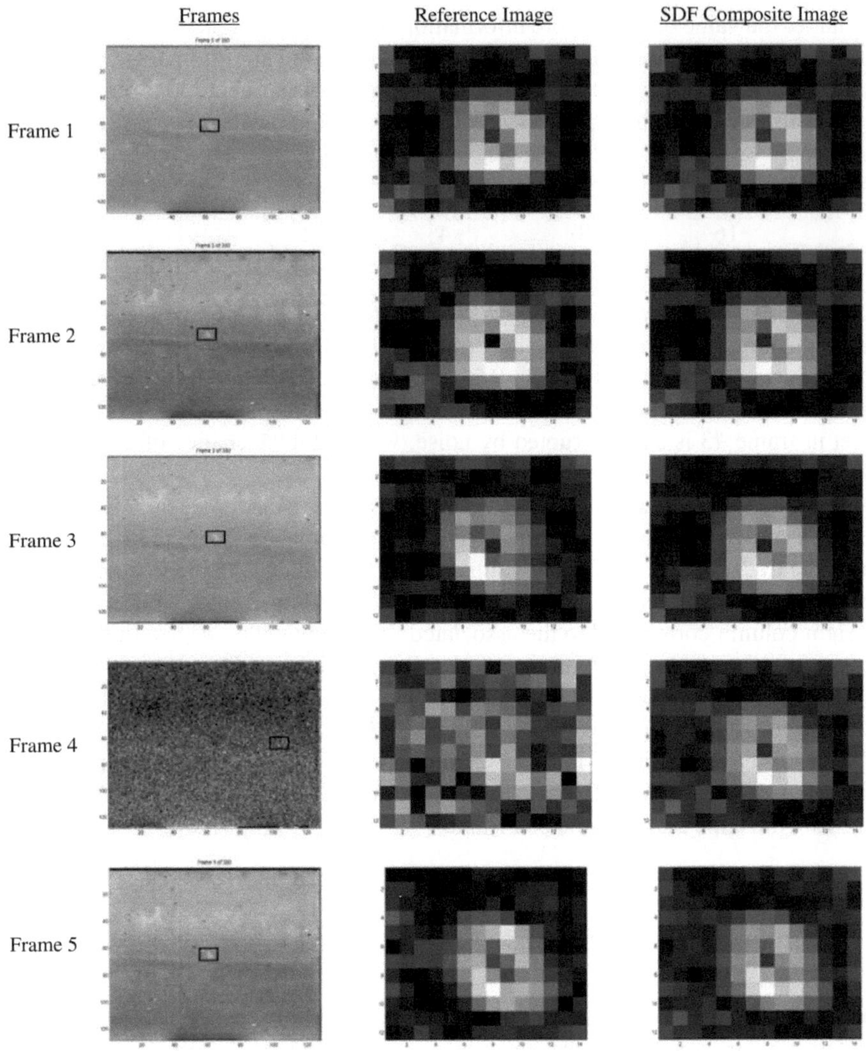

Fig. 15 Target modal update using the WCRF

the target. A multi-layered subimaging technique proposed in Ref. [38] is depicted in Fig. 16a, and requires that the target image size to be within 25% of the subimage size. It is also proposed to reduce the number of correlation operations by only correlating the subimages inside the radius of interest. Subimaging based segmentation approach [40] is inefficient because there are cases where none of the subimages may cover the entire image of the desired target as shown in Fig. 16b. In Ref. [30], a high-resolution image reconstruction technique is proposed by utilizing a number of shifted low-resolution subframes. By utilizing the reverse procedure, a frame available from a sequence is divided into a number of equal sized subframes. By shifting the grid in horizontal, vertical and diagonal directions, the entire target image can be captured in one of these subframes as shown in Figs. 16c–f, respectively. The size of each subframe is set equal to or greater than the size of the reference image, which is selected from a database or from a ground truth file. The entire input frame is divided into equal sized subframes using the aforementioned procedure, and the reference subframe is used for searching the unknown target in the entire frame where the subframe is a 25×25-pixel window. This method requires many iterations of correlation processing which are redundant because the target resides in only one of the segmented subframes. Besides, in high cluttered input scenes, it is difficult to select the accurate subframe, which contains the actual target.

In this section, we discuss an efficient way of utilizing the fringe-adjusted JTC algorithm for tracking the target. In this method, the tracking of a target in one frame is divided into two or more processes. The fringe-adjusted JTC based tracking is applied multiple times starting with the entire input scene as the input image with a large reference image obtained by segmentation from the preceding frame. Then the input image size and reference image size are gradually reduced until the reference image size becomes equal to the target size. In this way, unnecessary fringe-adjusted JTC processing is eliminated. This technique also effectively removes camera ego-motion. Multiple epoch processing not only compensates ego-motion, but also ensures efficient target tracking under complicated scenarios such as multiple independently moving objects, distortions due to skew, scale and rotation variations, as well as other artifacts.

Figure 17 shows the block diagram of the proposed fringe-adjusted JTC algorithm for target detection and tracking in a FLIR sequence. The proposed technique takes two consecutive frames from the image sequence and the location of the target in the previous frame i.e., reference frame as inputs and yields the location of the target in the subsequent frame as the output. The process is then repeated with the reference frame substituted by the preceding frame and the input frame substituted by the subsequent frame in the sequence.

In the first epoch of the process, the size of the input scene is set equal to the size of input frame while the reference image size is set equal to the reference image size or to a size larger than the target size. If the reference image size is very small, it may result in low peak-to-side lobe ratio in the fringe-adjusted JTC correlation output and may not be able to compensate high ego motion of the input image. Large reference image is favored but requires more epochs of the fringe-adjusted JTC process. The reference image size and the input image size are reduced in every epoch until the reference image size reaches the target size. The total number of epochs or iterations of the fringe-adjusted JTC process (m) depend on the tracking application, especially the processing speed. If the processing speed for running one epoch is faster than the output required by the application, additional epochs are preferred because the higher the number of epochs, the better the performance of the tracking algorithm.

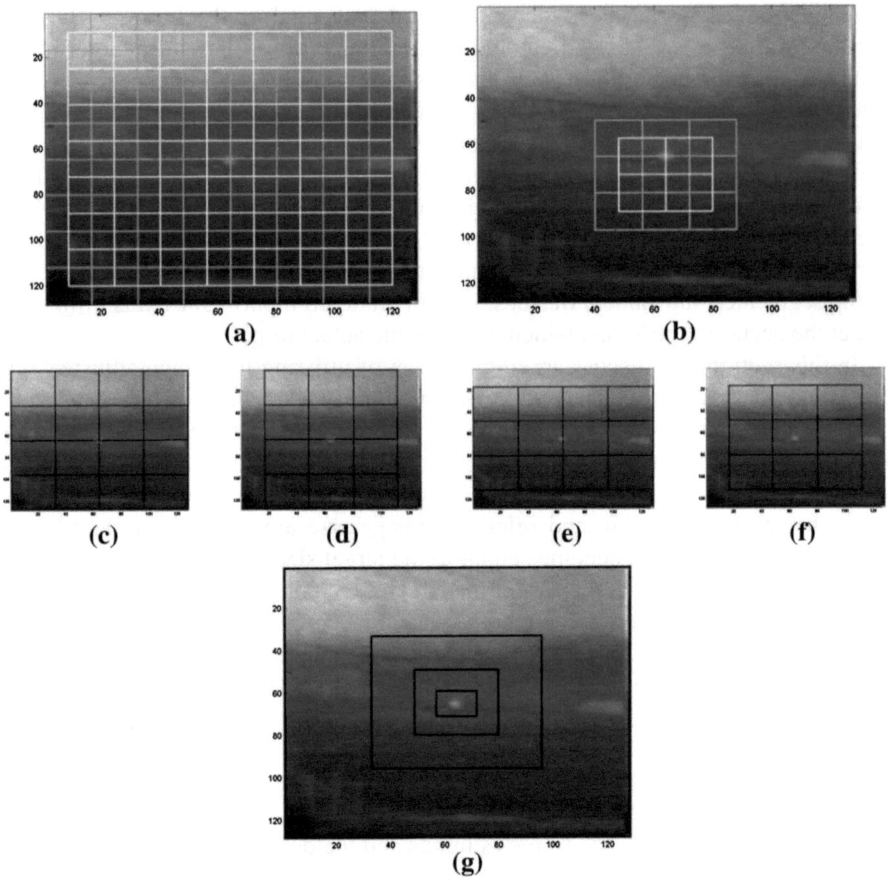

Fig. 16 Image segmentation: **a** two-layer sub-image grid [14], **b** two-layer sub-image grid with radius of interest module, **c–f** 49 (32 × 32) tracking regions are divided for a single (128 × 128) frame by conventional grid segmentation, **g** four tracking regions are divided for a single (128 × 128) frame by zoom segmentation

4.4 Test Results

The FLIR images used in the test involves hundreds of 128 × 128-pixel gray-level images in each sequence. Figure 18 illustrates fringe-adjusted JTC based detection and tracking results of a target in a FLIR sequence. In this example frame, a very small and low contrast tank is stationed at the junction of a road when the images are taken from an infrared camera installed on a moving platform. Figure 18a represents the input joint image of the first tracking process containing the entire preprocessed input image and the reference image segmented from the previous frame. Figure 18b depicts the correlation output produced by fringe-adjusted JTC corresponding to the

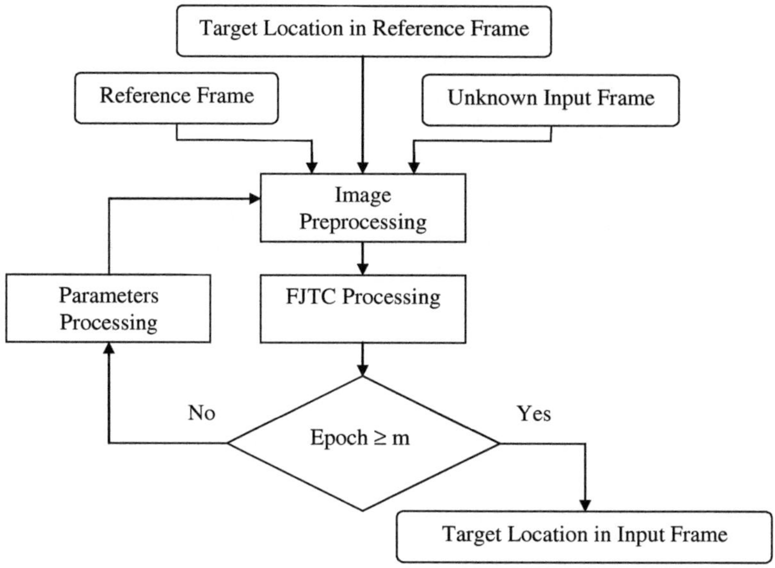

Fig. 17 Real-time object tracking using fringe-adjusted JTC algorithm

input joint image of Fig. 18a. Figure 18c illustrates the tracking result of the first iteration process indicating the best-fitted location of the reference image in the input scene. The outlined box in Fig. 18c indicates the position and size of the input image for the next fringe-adjusted JTC iteration process. Figure 18d represents the input joint image of the final tracking process of the frame while Fig. 18e shows the corresponding correlation output. Figure 18f depicts the tracking result obtained indicating the final position of the target.

Figure 19 demonstrated the tracking results of the fringe-adjusted JTC tracking algorithm for five challenging scenarios. The FLIR sequence in Fig. 19a and b illustrate the zoom in of two cold stationary tanks. The FLIR sequence in Fig. 19c illustrates the tracking of a tank in a low signal-to-noise ratio sequence. The FLIR sequence in Fig. 19d illustrates the tracking of a tank under high camera ego-motion. And the FLIR sequence in Fig. 19e illustrates the tracking of an independent moving object. The targets tracked by the fringe-adjusted JTC algorithm are shown by the rectangular black box. The target always remains inside the black box for the entire sequence indicating the effectiveness of the proposed tracking algorithm. Nevertheless, there are a few practical situations where the tracking algorithm may not function with 100% accuracy. Examples of such situations include overlapping of the target by other non-target objects, and rapid changes of temperature. One of the examples of overlapping target scenario is the tracking of tank behind a civilian vehicle. The WCRF model has the ability to mend some of the above mentioned scenarios since the updated model of the target contains feature information of the target from the

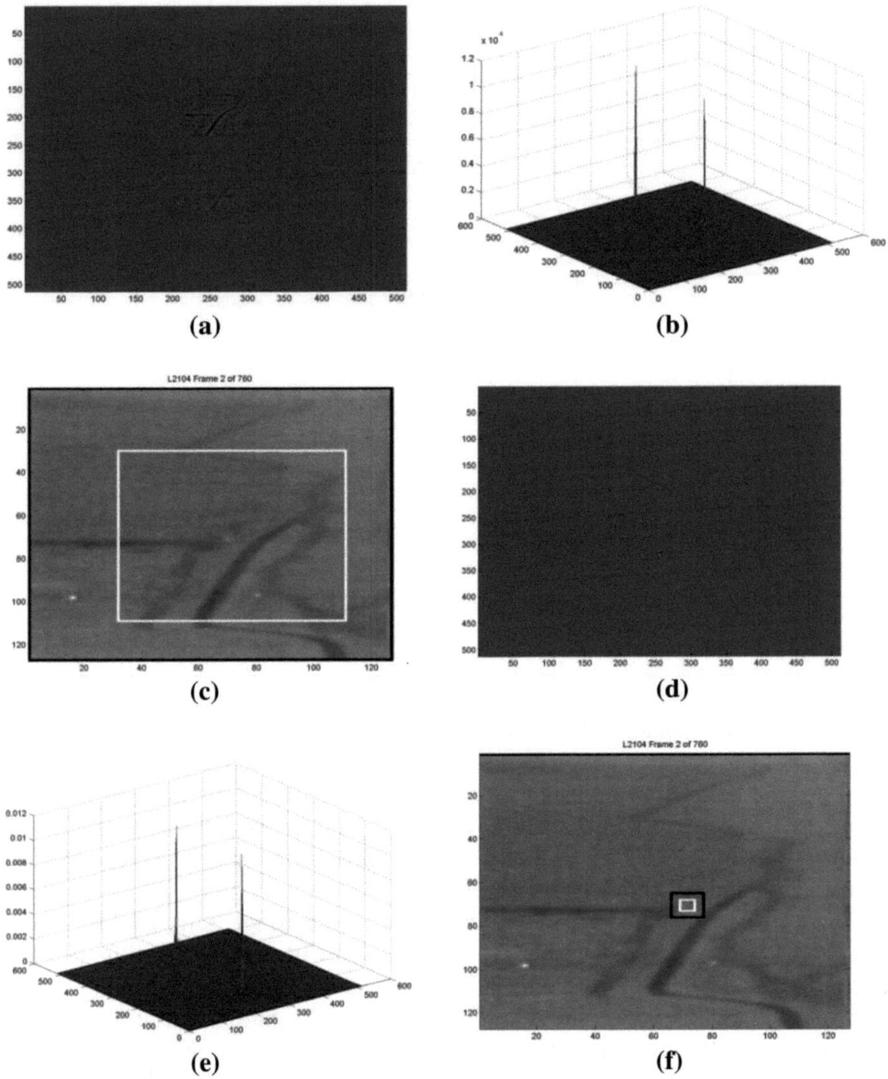

Fig. 18 Fringe-adjusted JTC based tracking of an almost invisible tank in one frame of a sequence: **a** input joint image corresponding to the first frame of the sequence, **b** correlation output obtained using the image of **a**, **c** target tracking using the result of **b**, **d** input joint image corresponding to the last frame of the sequence, **e** correlation output corresponding to **d**, and **f** target tracking using the result of **e** enclosed in the *rectangular white box* where the *rectangular black box* represents the input image of the last process

previous frames. Therefore, if the problem does not propagate for too many frames and the general appearance of the target does not change significantly, continuous tracking of the target using the fringe-adjusted JTC algorithm can still be achieved.

Fig. 19 Simulation results obtained using the proposed modified SDF based fringe-adjusted JTC tracking algorithm for four long wave and one medium wave FLIR sequences: **a** Frames #1, #50, #100 and #150 of the first sequence (L1803), **b** Frames #1, #100, #150, and #200 of the second sequence (L1902), **c** Frames #1, #100, #200 and #250 of the third sequence (L2008), **d** Frames #1, #200, #300, and #400 of the forth sequence (L2018), and **e** Frames #1, #100, #200, and #300 of the last sequence (M1413)

5 Conclusion

Real time recognition and tracking of targets in FLIR image sequences is a challenging problem due to low contrast, various 3D distortions, occlusion, overlapping as well as other artifacts. To alleviate these problems, a new distortion invariant real-time target tracking algorithm is proposed using a WCRF based fringe-adjusted JTC algorithm. This technique can track small objects comprising of only a few pixels while compensating the camera ego-motion. Simulation results using real life FLIR imagery verify the effectiveness of the proposed technique.

References

1. Alam, M.S., Haque, M., Khan, J., Kettani, H.: Target tracking in forward looking infrared imagery using fringe-adjusted joint transform correlation. Opt. Eng. **43**, 1407–1413 (2004)
2. Longmire, M.S., Takken, E.H.: LMS and matched digital filters for optical clutter suppression. Appl. Opt. **27**, 1141–1159 (1988)
3. Chen, J.Y., Reed, I.S.: A detection algorithm for optical targets in clutter. IEEE Trans. Aerosp. Electron. Syst. **23**, 46–59 (1987)
4. Bal, A., Alam, M.S.: Dynamic target tracking using fringe-adjusted joint transform correlation and template matching. Appl. Opt. **43**, 4874–4881 (2004)
5. Yilmaz, A., Shafique, K., Shah, M.: Target tracking in airborne forward looking infrared imagery. Image Vis. Comput. **21**, 623–635 (2003)
6. Loo, H.C., Alam, M.S.: Invariant object tracking using fringe-adjusted joint transform correlation. J. Opt. Eng. **43**, 2175–2183 (2004)
7. Bharadwaj, P., Carin, L.: Infrared-image classification using hidden Markov trees. IEEE Trans. Pattern Anal. Mach. Intell. **24**, 1394–1398 (2002)
8. Cooper, M.L., Miller, M.I.: Information measures for object recognition accommodating signature variability. IEEE Trans. Inf. Theory **46**, 1896–1907 (2000)
9. Alam, M.S., Bal, A.: Improved multiple target tracking via global motion compensation and optoelectronic correlation. IEEE Trans. Ind. Electron. **54**, 522–529 (2007)
10. Dawoud, A., Alam, M.S., Bal, A., Loo, C.: Decision fusion algorithm for target tracking in infrared imagery. Opt. Eng. **44**, 026401(1)–026401(8) (2005)
11. Lee, M., Kim, Y.: An efficient multitarget tracking algorithm for car applications. IEEE Trans. Ind. Electron. **50**, 397–400 (2003)
12. Nguyen, H.T., Smeulders, A.W.M.: Fast occluded object tracking by a robust appearance filter. IEEE Trans. Pattern Anal. Mach. Intell. **26**, 1099–1104 (2004)
13. Tao, H., Sawhney, H.S., Kumar, R.: Object tracking with Bayesian estimation of dynamic layer representations. IEEE Trans. Pattern Anal. Mach. Intell. **24**, 75–89 (2002)
14. Strehl, A., Aggarwal, J.K.: Detecting moving objects in airborne forward looking infra-red sequences. Mach. Vis. Appl. J. **11**, 267–276 (2000)
15. Mahalanobis, A., Sims, A.R., Nevel, A.V.: Signal-to-clutter measure for measuring automatic target recognition performance using complimentary eigenvalue distribution analysis. Opt. Eng. **42**, 1144–1151 (2003)
16. Mahalanobis, A., Muise, R.R., Stanfill, S.R., Nevel, A.V.: Design and application of quadratic correlation filters for target detection. IEEE Trans. Aerosp. Electron. Syst. **40**, 837–850 (2004)
17. Cheng, F., Yu, F.T.S., Gregory, D.A.: Multitarget detection using spatial synthesis joint transform correlator. Appl. Opt. **32**, 6521–6526 (1993)
18. Alam, M.S., Khan, J., Bal, A.: Hetero associative multiple target tracking using fringe-adjusted joint transform correlation. Appl. Opt. **43**, 328–365 (2004)
19. Miller, P.C., Royce, M., Virgo, P., Fiebig, M., Hamlyn, G.: Evaluation of an optical correlator automatic target recognition system for acquisition and tracking in densely cluttered natural scenes. Opt. Eng. **38**, 1814–1825 (1999)
20. Huo, X.: A statistical analysis of Fukunaga–Koontz transform. IEEE Signal Process. Lett. **11**, 123–126 (2004)

21. Castellano, G., Boyce, J., Sandler, M.: Regularized CDWT optical flow applied to moving-target detection in IR imagery. Mach. Vis. Appl. **11**, 277–288 (2000)
22. Dahyot, R., Charbonnier, P., Heitz, F.: Unsupervised statistical detection of changing objects in camera-in-motion video. In: Proceedings of the IEEE International Conference on Image Processing (ICIP'01), Greece, October (2001)
23. Elnagar, A., Basu, A.: Motion detection using background constraints. Pattern Recognition **28**, 1537–1554 (1995)
24. Bruno, M.G.S.: Sequential importance sampling filtering for target tracking in image sequences. IEEE Trans. Ind. Electron. **10**, 246–300 (2003)
25. Davies, D., Palmer, P., Mirmehdi, M.: Detection and tracking of very small low contrast objects. In: Ninth British Machine Vision Conference, September 1998 (1998)
26. VanderLugt, A.: Signal detection by complex spatial filtering. IEEE Trans. Inf. Theory **10**, 139–145 (1964)
27. Weaver, C.S., Goodman, J.W.: A technique for optically convolving two functions. Appl. Opt. **5**, 1246–1249 (1966)
28. Yu, F.T.S., Lu, X.J.: A real-time programmable joint transform correlator. Opt. Commun. **52**, 10–16 (1984)
29. Alam, M.S., Chain, D.: Efficient multiple target recognition using a wavelet transform processor. Opt. Eng. **39**, 1203–1210 (2000)
30. Alam, M.S., Karim, M.A.: Multiple target detection using a modified fringe-adjusted joint transform correlator. Opt. Eng. **33**, 1610–1617 (1994)
31. Alam, M.S.: Phase-encoded fringe-adjusted joint transform correlation. Opt. Eng. **39**, 1169–1176 (2000)
32. Briechle, K., Hanebeck, U.D.: Template matching using fast normalized cross correlation. In: Optical Pattern Recognition XII, Proceeding of SPIE, vol. 4387, pp. 95–102 (2001)
33. Javidi, B., Kuo, C.: Joint transform image correlation using a binary spatial light modulator at the Fourier plane. Appl. Opt. **27**, 663–665 (1988)
34. Alam, M.S.: Deblurring using fringe-adjusted joint transform correlation. Opt. Eng. **37**, 556–564 (1998)
35. Alam, M.S., Karim, M.A.: Improved correlation discrimination in a multiobject bipolar joint transform correlator. Opt. Laser Tech. **24**, 45–50 (1992)
36. Yu, F.T.S., Cheng, F., Nagata, T., Gregory, D.A.: Effects of fringe binarization of multi-object joint transform correlation. Appl. Opt. **28**, 2988–2990 (1989)
37. Wu, Y., Huang, T.S.: Non-stationary color tracking for vision-based human–computer interaction. IEEE Trans. Neural Network **13**, 948–960 (2002)
38. Oron, E., Kumar, A., Bar-Shalom, Y.: Precision tracking with segmentation for imaging sensor. IEEE Trans. Aerosp. Electron. Syst. **29**, 977–987 (1993)
39. Rastogi, K., Chatterji, B.N., Ray, A.K.: Design of real-time tracking system for fast moving objects. IETE J. Res. **43**, 359–369 (1997)
40. Gudmundsson, K., Awwal, A.A.S.: Sub-imaging technique to improve phase only filter search capability. Appl. Opt. **42**, 4709–4717 (2003)
41. Alam, M.S., Bognar, J.G., Hardie, R.C., Yasuda, B.J.: Infrared image registration and high resolution reconstruction using multiple translationally shifted aliased video frames. IEEE Trans. Instrum. Meas. **49**, 915–923 (2000)
42. Bal, A., Alam, M.S.: Automatic target tracking in FLIR image sequences. In: Proceedings of the SPIE Conference on Automatic Target Recognition XIV, vol. 5426, pp. 30–36 (2004)
43. Shekarforoush, H., Chellappa, R.: A multi-fractal formalism for stabilization, object detection and tracking in FLIR sequences. In: IEEE International Conference on Image Processing, vol. 3, pp. 78–81 (2000)
44. Braga-Neto, U., Goutsias, J.: Automatic target detection and tracking in forward-looking infrared image sequences using morphological connected operators. In: 33rd Conference of Information Sciences and Systems, vol. 1, pp. 173–178, March 1999 (1999)

45. Comaniciu, D., Ramesh, V., Meer, P.: Real-time tracking of non-rigid objects using mean shift. In: IEEE Conference on Computer Vision and Pattern Recognition, vol. 2, pp. 142–149 (2000)
46. Feng, D., Zhao, H., Xia, S.: Amplitude-modulated JTC for improving correlation discrimination. Opt. Commun. **86**, 260–264 (1991)
47. Alam, M.S., Karim, M.A.: Fringe-adjusted joint transform correlation. Appl. Opt. **32**(23), 4344–4350 (1993)
48. Alam, M.S., Chen, X.W., Karim, M.A.: Distortion-invariant fringe-adjusted joint transform correlation. Appl. Opt. **36**, 7422–7427 (1997)
49. Grycewicz, T.J.: Applying time modulation to the joint transform correlator. Opt. Eng. **33**(6), 1813–1830 (1994)
50. Casasent, D., Chang, W.T.: Correlation synthetic discriminant functions. Appl. Opt. **25**, 2343–2350 (1986)
51. Wu, Y., Huang, T.S.: Nonstationary color tracking for vision-based human–computer interaction. IEEE Trans. Neural Networks **13**, 948–960 (2002)
52. Alam, M.S., et al.: Fringe-adjusted JTC based target detection and tracking using subframes from a video sequence. In: Proceedings of the SPIE, vol. 5201, pp. 85–96, San Diego, 3–8 August 2003

A Bayesian Method for Infrared Face Recognition

Tarek Elguebaly and Nizar Bouguila

Abstract In the context of face recognition, an important problem is accurate identification under variable illumination conditions. This problem has received relatively more attention in visible spectrum domain compared to the thermal infrared one. This was justified by both the higher cost of thermal sensors, the lack of widely available IR image databases and the quality of the produced images (lower resolution and higher image noise). Recently, thermal imagery of human faces has been established as a valid biometric signature and several approaches have been proposed, to tackle the problem of infrared face recognition, thanks to the advances of infrared imaging technology [Prokoski, Proceedings of the IEEE Workshop on Computer Vision Beyond the Visible Spectrum: Methods and Applications 5–14, 2000]. Some of these approaches have been based on machine learning techniques by supposing that the extracted infrared face features are Gaussian which is not generally an appropriate assumption. Motivated by the fact that infrared images are generally characterized by non-Gaussian features impossible to model using rigid distributions such as the Gaussian, we propose, in this chapter, an efficient Bayesian unsupervised algorithm for infrared face recognition, based on the Generalized Gaussian mixture model.

Keywords Thermal infrared imaging · Face recognition · Edge direction Histogram · Generalized Gaussian distribution · Mixture modeling · Bayesian analysis · Metropolis-Hastings · Gibbs sampling.

T. Elguebaly (✉) · N. Bouguila
Concordia Institute for Information Systems Engineering (CIISE),
Concordia University, Montreal, QC H3G 1T7, Canada
e-mail: t_elgue@encs.concordia.ca

N. Bouguila
e-mail: bouguila@ciise.concordia.ca

R. Hammoud et al. (eds.), *Machine Vision Beyond Visible Spectrum*,
Augmented Vision and Reality, 1, DOI: 10.1007/978-3-642-11568-4_5,

1 Introduction

In recent years, face recognition has become one of the most rapidly growing research areas in computer vision, due to its wide range of potential applications related to security and safety. Moreover, face recognition is generally a key component for smart environments. Face recognition is actually the most natural intuitive way to identify individuals compared to several others biometric authentication methods such as fingerprints, iris patterns and voice print that generally rely on the cooperation of the participants. For these reasons, applications for face recognition technology are booming in areas where personal identification is critical (e.g., surveillance, access control, etc.) [35]. Face recognition can be viewed as the process that given an unknown input face, the system reveals its identity by comparing it with a database of known persons. This process is generally based on three main tasks: face detection, feature extraction, and face identification. Face detection is used to identify face-like objects from different other objects. In order to decrease the dimensionality of face images, they are usually represented in terms of low-level feature vectors in lower dimensional feature space (i.e., face signature) for recognition. Face identification task identifies the input person face by searching a database of known individuals. Despite the variety of approaches and tools studied, face recognition is still a challenging problem due to the variations in appearance that a given face may have in a scene and because of the different image acquisition problems such as noise, video–camera distortion, and image resolution [35]. This is especially the case of face recognition in visible-spectrum which has received huge attention in the literature. Performance of visual face recognition is sensitive to variations in illumination conditions, even more than variations raised from changes in face identity [42]. Other factors that further complicate the face recognition task are facial expressions [44] and pose variations [4].

Recently, different studies have shown that thermal IR offers a promising alternative to visible imagery for handling variations in face appearance [7, 9, 43] due to illumination changes [22], facial expressions [9, 12], and face poses [12]. A comparison of visible and infrared imagery for face recognition can be found in [16]. Thermal IR imagery offers a capability for identification under different lighting conditions including total darkness [2]. Thus, IR-based algorithms have the potential to provide simpler and more robust solutions, improving performance in uncontrolled environments and deliberate attempts to obscure identity [13]. Figure 1 shows an example of visual and thermal face image characteristics. From this figure we can see clearly that thermal characteristics of the face are not sensitive to variations in illumination and facial expression which have actually changed the visual appearance of the face. Several approaches have been proposed to analyze and recognize infrared faces and can be divided into two main groups: appearance-based and feature-based methods. While appearance-based methods focus on the global properties of the face, feature-based methods explore the facial features (e.g., eyes, mouth) statistical and geometrical properties [11, 32]. For instance, facial expression recognition based on thermal imaging has been investigated in [43]. In [33] the authors have

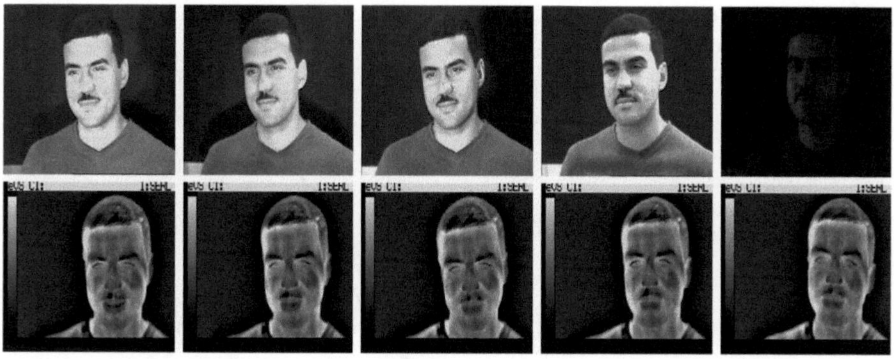

Fig. 1 Visual and thermal image characteristics of faces with variations in illumination

proposed a two-stage face recognition method based on Gabor filtering and Bessel modeling. A PCA-based approach has been proposed in [41]. The authors in [8] have studied thermal face recognition over time. An unsupervised local and global feature extraction paradigm has been proposed in [21] to approach the problem of face expression recognition in thermal images. In [32] the authors observed that visual and thermal spectra modalities are complementary and have proposed a face recognition approach by fusing the two. Moreover, many machine learning techniques have been proposed. A comprehensive review of such techniques can be found in [35]. Many of these approaches, however, suppose that the extracted infrared face features are Gaussian which is not generally an appropriate assumption. We propose then, in this chapter, an appearance-based approach using unsupervised Bayesian learning of finite generalized Gaussian mixture models.

The rest of this chapter is organized as follows. Section 2 presents the finite generalized Gaussian mixture (GGM) in details. In Sect. 3, the complete Bayesian learning algorithm for the GGM is presented. The experimental results are given in Sect. 4. Our last section is devoted to the conclusion.

2 Generalized Gaussian Mixture Model

Mixture models are one of the machine learning techniques receiving considerable attention in different applications. Mixture models are normally used to model complex datasets. In most of image processing and computer vision applications, the Gaussian density is applied for data modeling. However, in the majority of data generated from computer vision applications are non-Gaussian. Many studies have demonstrated that the Generalized Gaussian Distribution (GGD) can be a good alternative to the Gaussian thanks to its shape flexibility which allows the modeling of a large number of non-Gaussian signals [15, 31, 39, 45]. The one-dimensional GGD for a variable $x \in \mathbb{R}$ is defined as follows [20, 36]:

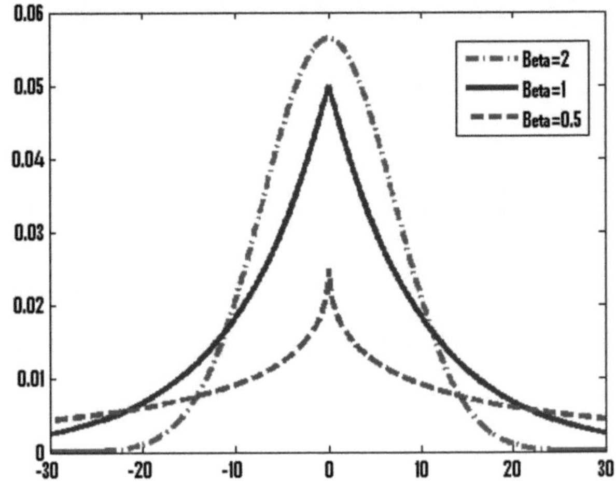

Fig. 2 Generalized Gaussian distributions with different values of the shape parameter

$$P(x|\mu,\alpha,\beta) = \frac{\beta\alpha}{2\Gamma(1/\beta)} e^{-(\alpha|x-\mu|)^{\beta}} \tag{1}$$

where $\alpha = \frac{1}{\sigma}\sqrt{\frac{\Gamma(3/\beta)}{\Gamma(1/\beta)}}$, $-\infty < \mu < \infty, \sigma > 0$ is the standard deviation, $\beta > 0$, and $\alpha > 0$, and $\Gamma(\cdot)$ is the Gamma function given by: $\Gamma(x) = \int_0^{\infty} t^{x-1}e^{-t}\,dt,\ x > 0$. μ, α and β denote the distribution mean, the inverse scale parameter, and the shape parameter, respectively. The GGD is flexible thanks to its shape parameter β that controls the decay rate of the density function. In other words, β allows the GGD to take different shapes depending on the data. Figure 2 shows us two main reasons to use GGD. First, the parameter β controls the shape of the pdf. The larger the value, the flatter the pdf; and the smaller the value, the more picked the pdf. Second, when $\beta = 2$ and $\beta = 1$, the GGD is reduced to the Gaussian and Laplacian distributions, respectively.

Having a d-dimensional vector $\mathbf{X} = (X_1, \ldots, X_d)$, where each element follows a GGD, its probability density function can be expressed as follows:

$$P(\mathbf{X}|\mu,\alpha,\beta) = \prod_{k=1}^{d} \frac{\beta_k\alpha_k}{2\Gamma(1/\beta_k)} e^{-(\alpha_k|X_k-\mu_k|)^{\beta_k}} \tag{2}$$

where $\mu = (\mu_1, \ldots, \mu_d), \alpha = (\alpha_1, \ldots, \alpha_d)$, and $\beta = (\beta_1, \ldots, \beta_d)$ are the mean, the inverse scale, and the shape parameters of the d-dimensional GGD, respectively.

Let $\mathscr{X} = (\mathbf{X}_1, \ldots, \mathbf{X}_N)$ be a set of N independent and identical distributed vectors assumed to arise from a finite General Gaussian mixture model with M components:

$$P(\mathbf{X}|\Theta) = \sum_{j=1}^{M} p_j P(\mathbf{X}|\mu_j, \alpha_j, \beta_j) \tag{3}$$

where p_j are the mixing proportions which must be positive and sum to one, $j = 1, \dots, M$, are the conditional probabilities. The set of parameters of the mixture with M classes is defined by $\Theta = (\mu_1, \dots, \mu_M, \alpha_1, \dots, \alpha_M, \beta_1, \dots, \beta_M, p)$, $p = (p_1, \dots, p_M)$.

In the past few years, several approaches have been applied for GGD's parameters estimation such as moment estimation [20, 23], entropy matching estimation [3, 19], and maximum likelihood estimation [10, 23–25, 39]. It is noteworthy that these approaches consider a single distribution. Concerning finite mixture models parameters estimation, approaches can be arranged into two categories: deterministic and Bayesian methods. In deterministic approaches, parameters are taken as fixed and unknown, and inference is founded on the likelihood of the data. In the recent past, some deterministic approaches have been proposed for the estimation of finite generalized Gaussian mixture (GGM) models parameters (see, for instance, [26–29, 38]). Despite the fact that deterministic approaches have controlled mixture models estimation due to their small computational time, many works have demonstrated that these methods have severe problems such as convergence to local maxima, and their tendency to over fit the data [5] especially when data are sparse or noisy. With the computational tools evolution, researchers were encouraged to implement and use Bayesian MCMC methods and techniques as an alternative approach. Bayesian methods consider parameters to be random, and to follow different probability distributions (prior distributions). These distributions are used to describe our knowledge before considering the data, as for updating our prior beliefs the likelihood is used. Refer to [5, 18] for interesting and in depth discussions about the general Bayesian theory. The development of methods that precisely incorporate prior knowledge and uncertainty into the decision-making process are of a huge significance to the computer vision field.

3 Bayesian Learning of the GGM

We introduce stochastic indicator vectors, $Z_i = (Z_{i1}, \dots, Z_{iM})$, one for each observation, whose role is to encode to which component the observation belongs. In other words, Z_{ij}, the unobserved or missing vector, equals 1 if $\mathbf{X}_i$ belongs to class j and 0, otherwise. The complete-data likelihood for this case is then:

$$P(\mathcal{X}, Z|\Theta) = \prod_{i=1}^{N} \prod_{j=1}^{M} (P(\mathbf{X}_i|\xi_j)p_j)^{Z_{ij}} \tag{4}$$

where $Z = \{Z_1, Z_2, \dots, Z_N\}$, and $\xi_j = (\mu_j, \alpha_j, \beta_j)$. For the moment, we suppose the number of mixture M is known. Bayesian MCMC simulation methods are found on the Bayesian theory, which means that they allow for probability statements to be made directly about the unknown parameters of a mixture model, while taking into consideration prior or expert opinion. The well-known Bayesian formula is:

$$\pi(\Theta|\mathscr{X}, Z) = \frac{\pi(\Theta)P(\mathscr{X}, Z|\Theta)}{\int \pi(\Theta)P(\mathscr{X}, Z|\Theta)} \propto \pi(\Theta)P(\mathscr{X}, Z|\Theta) \tag{5}$$

where $(\mathscr{X}, Z)$, $\pi(\Theta)$, $P(\mathscr{X}, Z|\Theta)$, and $\pi(\Theta|\mathscr{X}, Z)$ are the complete data, the prior information about the parameters, the realization of the complete data, and the posterior distribution, respectively. This means that in order to get the posterior distribution using MCMC, we need to combine the prior information about the parameters, $\pi(\Theta)$, with the observed value or realization of the complete data $P(\mathscr{X}, Z|\Theta)$. With the joint distribution, $\pi(\Theta)P(\mathscr{X}, Z|\Theta)$, in hand we can deduce the posterior distribution (Eq. 5). Having $\pi(\Theta|\mathscr{X}, Z)$ we can simulate our model parameters Θ. In the following section, the prior for the parameters will be specified, and the conditional distributions for these will be derived.

3.1 Hierarchical Model, Priors and Posteriors

In this section, the unknown parameters, Θ, in our mixture model are regarded as random variables drawn from some prior distributions. We simulate Z_i according to the posterior probability $\pi(Z_i|\Theta, \mathscr{X})$, chosen to be Multinomial of order one with a weight given by $\widehat{Z}_{ij}(\mathscr{M}(1; \widehat{Z}_{i1}, \ldots, \widehat{Z}_{iM}))$. This choice is due to two reasons, first, we know that each Z_i is a vector of zero-one indicator variables to define from which component j the $\mathbf{X}_i$ observation arises. Second, the probability that the ith observation, $\mathbf{X}_i$, arises from the jth component of the mixture is given by $\widehat{Z}_{ij}$ where

$$\widehat{Z}_{ij} = \frac{p_j P(\mathbf{X}_i|\mu_j, \alpha_j, \beta_j)}{\sum_{j=1}^{M} p_j P(\mathbf{X}_i|\mu_j, \alpha_j, \beta_j)} \tag{6}$$

Now to simulate p we need to get $\pi(p|Z)$, using Bayes rule:

$$\pi(p|Z) = \frac{\pi(Z|p)\pi(p)}{\int \pi(Z|p)\pi(p)} \propto \pi(Z|p)\pi(p) \tag{7}$$

This means that we need to determine $\pi(Z|p)$ and $\pi(p)$. It is well known that the vector p is defined as ($\sum_{j=1}^{M} p_j = 1$, where $p_j \geq 0$), then the commonly considered choice as a prior is the Dirichlet distribution [5, 18]:

$$\pi(p) = \frac{\Gamma(\sum_{j=1}^{M} \eta_j)}{\prod_{j=1}^{M} \Gamma(\eta_j)} \prod_{j=1}^{M} p_j^{\eta_j - 1} \tag{8}$$

where $(\eta_1, \ldots, \eta_M)$ is the parameters vector of the Dirichlet distribution. As for $\pi(Z|p)$ we have:

$$\pi(Z|p) = \prod_{i=1}^{N} \prod_{j=1}^{M} p_j^{Z_{ij}} = \prod_{j=1}^{M} p_j^{n_j} \tag{9}$$

where $n_j = \sum_{i=1}^{N} \boldsymbol{I}_{Z_{ij=1}}$, then we can conclude that:

$$\pi(p|Z) = \pi(Z|P)\pi(p) = \frac{\Gamma(\sum_{j=1}^{M}\eta_j)}{\prod_{j=1}^{M}\Gamma(\eta_j)} \prod_{j=1}^{M} p_j^{\eta_j+n_j-1} \propto \mathscr{D}(\eta_1+n_1, \ldots, \eta_M+n_M) \tag{10}$$

where $\mathscr{D}$ denotes the Dirichlet distribution with parameters $(\eta_1+n_1, \ldots, \eta_M+n_M)$. Thus, we can deduce that the Dirichlet distribution is a conjugate prior for the mixture proportions (i.e., the prior and the posterior have the same form).

For the parameters ξ, we assigned independent Normal priors for the distributions means (μ_j) with common hyper parameters δ and ε^2 as the mean and variance, respectively. Independent Gamma priors with common hyper parameters ι and ρ as the shape and rate parameters, respectively, are assigned for the inverse scale α_j. For the shape parameters, β_j, we used independent Gamma priors with common hyper parameters κ and ς as the shape and rate parameters, respectively [6]:

$$P(\mu_j|\delta, \varepsilon^2) \sim \prod_{k=1}^{d} \frac{1}{\sqrt{2\pi}\varepsilon} \mathbf{e}^{\frac{-(\mu_{jk}-\delta)^2}{2\varepsilon^2}} \tag{11}$$

$$P(\alpha_j|\iota, \rho) \sim \prod_{k=1}^{d} \frac{\alpha_{jk}^{\iota-1}\rho^{\iota}\mathbf{e}^{-\rho\alpha_{jk}}}{\Gamma(\iota)} \tag{12}$$

$$P(\beta_j|\kappa, \varsigma) \sim \prod_{k=1}^{d} \frac{\beta_{jk}^{\kappa-1}\varsigma^{\kappa}\mathbf{e}^{-\varsigma\beta_{jk}}}{\Gamma(\kappa)} \tag{13}$$

δ, ε^2, ι, ρ, κ, ς are called the hyperparameters of the model. After selecting the parameters priors, we can deduce the posterior distributions for μ_j, α_j, and β_j:

$$\begin{aligned} P(\mu_j|Z, \mathscr{X}) &\propto P(\mu_j|\delta, \varepsilon^2) \prod_{Z_{ij=1}} P(\mathscr{X}|\mu_j, \alpha_j, \beta_j) \\ &\propto \prod_{k=1}^{d} \frac{1}{\varepsilon} \mathbf{e}^{\frac{-(\mu_{jk}-\delta)^2}{2\varepsilon^2}} \times \prod_{k=1}^{d} \mathbf{e}^{\sum_{Z_{ij=1}}(-\alpha_{jk}|X_{ik}-\mu_{jk}|)^{\beta_{jk}}} \end{aligned} \tag{14}$$

$$\begin{aligned} P(\alpha_j|Z, \mathscr{X}) &\propto P(\alpha_j|\iota, \rho) \prod_{Z_{ij=1}} P(\mathscr{X}|\mu_j, \alpha_j, \beta_j) \\ &\propto \prod_{k=1}^{d} \frac{\alpha_{jk}^{\iota-1}\rho^{\iota}\mathbf{e}^{-\rho\alpha_{jk}}}{\Gamma(\iota)} \times \prod_{k=1}^{d} [\alpha_{jk}]^{n_j} \mathbf{e}^{\sum_{Z_{ij=1}}(-\alpha_{jk}|X_{ik}-\mu_{jk}|)^{\beta_{jk}}} \end{aligned} \tag{15}$$

$$\begin{aligned} P(\beta_j|Z, \mathscr{X}) &\propto P(\beta_j|\kappa, \varsigma) \prod_{Z_{ij=1}} P(\mathscr{X}|\mu_j, \alpha_j, \beta_j) \\ &\propto \prod_{k=1}^{d} \frac{\beta_{jk}^{\kappa-1}\varsigma^{\kappa}\mathbf{e}^{-\varsigma\beta_{jk}}}{\Gamma(\kappa)} \times \prod_{k=1}^{d} \left[\frac{\beta_{jk}}{\Gamma(1/\beta_{jk})}\right]^{n_j} \mathbf{e}^{\sum_{Z_{ij=1}}(-\alpha_{jk}|X_{ik}-\mu_{jk}|)^{\beta_{jk}}} \end{aligned} \tag{16}$$

In order to have a more flexible model, we introduce an additional hierarchical level by allowing the hyperparameters to follow some selected distributions. The hyperparameters δ and ε^2 associated with the μ_j are given Normal and inverse Gamma priors, respectively:

$$P(\delta|\varepsilon,\chi^2) \sim \frac{1}{\sqrt{2\pi}\chi}\mathbf{e}^{\frac{-(\delta-\varepsilon)^2}{2\chi^2}} \tag{17}$$

$$P(\varepsilon^2|\varphi,\rho) \sim \frac{\rho^{\varphi}\mathbf{e}^{(-\rho/\varepsilon^2)}}{\Gamma(\varphi)\varepsilon^{2(\varphi+1)}} \tag{18}$$

Thus, according to these two previous equations and Eq. 11, we have

$$P(\delta|\ldots) \propto P(\delta|\varepsilon,\chi^2)\prod_{j=1}^{M} P(\mu_j|\delta,\varepsilon^2) \propto \mathbf{e}^{\frac{-(\delta-\varepsilon)^2}{2\chi^2}} \times \prod_{j=1}^{M}\prod_{k=1}^{d}\mathbf{e}^{\frac{-(\mu_{jk}-\delta)^2}{2\varepsilon^2}} \tag{19}$$

$$\begin{aligned} P(\varepsilon^2|\ldots) &\propto P(\varepsilon^2|\varphi,\rho)\prod_{j=1}^{M} P(\mu_j|\delta,\varepsilon^2) \\ &\propto \frac{\exp(-\rho/\varepsilon^2)}{\varepsilon^{2(\varphi+1)}}\left[\frac{1}{\varepsilon}\right]^{Md} \times \prod_{j=1}^{M}\prod_{k=1}^{d}\mathbf{e}^{\frac{-(\mu_{jk}-\delta)^2}{2\varepsilon^2}} \end{aligned} \tag{20}$$

The hyperparameters ι and ρ associated with the α_j are given inverse Gamma and Gamma priors, respectively:

$$P(\iota|\vartheta,\varpi) \sim \frac{\varpi^{\vartheta}\mathbf{e}^{(-\varpi/\iota)}}{\Gamma(\vartheta)\iota^{\vartheta+1}} \tag{21}$$

$$P(\rho|\tau,\omega) \sim \frac{\rho^{\tau-1}\omega^{\tau}\mathbf{e}^{-\omega\rho}}{\Gamma(\tau)} \tag{22}$$

Thus, according to these two previous equations and Eq. 12, we have

$$P(\iota|\ldots) \propto P(\alpha_\alpha|\vartheta,\varpi)\prod_{j=1}^{M} P(\alpha_j|\iota,\rho) \propto \frac{\mathbf{e}^{(-\varpi/\iota)}}{\iota^{\vartheta+1}}\left[\frac{\rho^{\iota}}{\Gamma(\iota)}\right]^{Md} \times \prod_{j=1}^{M}\prod_{k=1}^{d}\alpha_{jk}^{\iota-1}\mathbf{e}^{-\rho\alpha_{jk}} \tag{23}$$

$$P(\rho|\ldots) \propto P(\beta_\alpha|\tau,\omega)\prod_{j=1}^{M} P(\alpha_j|\iota,\rho) \propto \rho^{\tau-1}\mathbf{e}^{-\omega\rho}\left[\rho^{\iota}\right]^{Md} \times \prod_{j=1}^{M}\prod_{k=1}^{d}\alpha_{jk}^{\iota-1}\mathbf{e}^{-\rho\alpha_{jk}} \tag{24}$$

The hyperparameters κ and ς associated with the β_j are given inverse Gamma and Gamma priors, respectively:

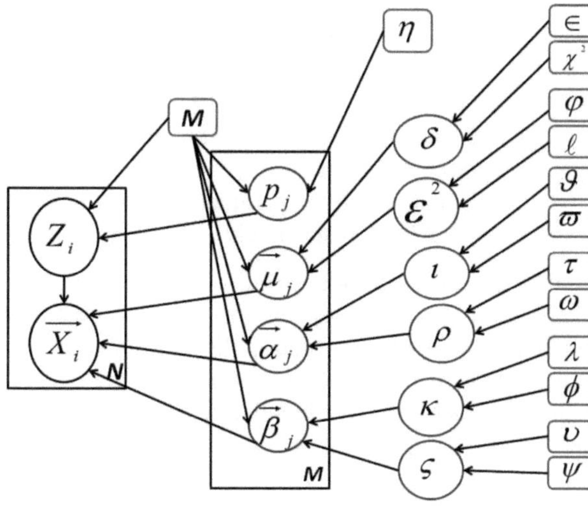

Fig. 3 Graphical representation of the Bayesian hierarchical finite general Gaussian mixture model. *Nodes* in this graph represent random variables, *rounded boxes* are fixed hyperparameters, *boxes* indicate repetition (with the number of repetitions in the *lower right*) and *arcs* describe conditional dependencies between variables

$$P(\kappa|\lambda,\phi) \sim \frac{\phi^{\lambda}\mathbf{e}^{(-\phi/\kappa)}}{\Gamma(\lambda)\kappa^{\lambda+1}} \tag{25}$$

$$P(\varsigma|\nu,\psi) \sim \frac{\varsigma^{\nu-1}\psi^{\nu}\mathbf{e}^{-\psi\varsigma}}{\Gamma(\nu)} \tag{26}$$

Thus, according to these two previous equations and Eq. 13, we have

$$P(\kappa|\ldots) \propto P(\kappa|\lambda,\phi)\prod_{j=1}^{M} P(\beta_j|\kappa,\varsigma) \propto \frac{\mathbf{e}^{(-\phi/\kappa)}}{\kappa^{\lambda+1}}\left[\frac{\varsigma^{\kappa}}{\Gamma(\kappa)}\right]^{Md} \times \prod_{j=1}^{M}\prod_{k=1}^{d}\beta_{jk}^{\kappa-1}\mathbf{e}^{-\varsigma\beta_{jk}} \tag{27}$$

$$P(\varsigma|\ldots) \propto P(\varsigma|\nu,\psi)\prod_{j=1}^{M} P(\beta_j|\kappa,\varsigma) \propto \varsigma^{\nu-1}\mathbf{e}^{-\psi\varsigma}\left[\varsigma^{\kappa}\right]^{Md} \times \prod_{j=1}^{M}\prod_{k=1}^{d}\beta_{jk}^{\kappa-1}\mathbf{e}^{-\varsigma\beta_{jk}} \tag{28}$$

Our hierarchical model can be displayed as a directed acyclic graph (DAG) as shown in Fig. 3.

3.2 Complete Algorithm

Having all the conditional posteriors, we can employ a Gibbs sampler [37] with steps as follow:

1. Initialization of the model parameters.
2. Step t, for $t = 1, \ldots$

(a) Generate $Z_i^{(t)} \sim \mathscr{M}(1; \widehat{Z}_{i1}, \ldots, \widehat{Z}_{iM})$
(b) Compute $n_j^{(t)} = \sum_{i=1}^{N} \mathbb{I}_{Z_{ij}^{(t)}}$
(c) Generate $p^{(t)}$ from Eq. 10.
(d) Generate the mixture parameters μ_j, α_j, and β_j from Eqs. 14, 15, 16, respectively, for $j = 1, \ldots, M$.
(e) Update the hyperparameters δ, ε^2, ι, ρ, κ, and ς from Eqs. 19, 20, 23, 24, 27, 28, respectively.

It is quite easy to notice that we cannot simulate directly from these posterior distributions because they are not in well known forms. To solve this problem we applied the well known Metropolis–Hastings (M–H) algorithm given in [40]. The major problem in the M–H algorithm is the choice of the proposal distribution. Random walk M–H given in [40] is used here to solve this problem, then the proposals are considered to be: $\widetilde{\mu_{jk}} \sim \mathscr{N}(\mu_{jk}^{(t-1)}, \zeta^2)$, $\widetilde{\alpha_{jk}} \sim \mathscr{L}\mathscr{N}(\log(\alpha_{jk}^{(t-1)}), \zeta^2)$, $\widetilde{\beta_{jk}} \sim \mathscr{L}\mathscr{N}(\log(\beta_{jk}^{(t-1)}), \zeta^2)$, where $k = (1, \ldots, d)$, $\mathscr{L}\mathscr{N}$ is the log-normal distribution, since, we know that $\widetilde{\alpha_j} > 0$ and $\widetilde{\beta_j} > 0$. ζ^2 is the scale of the random walk.

It is noteworthy that choosing a relevant model consists both of choosing its form and the number of components M. The integrated or marginal likelihood using the Laplace–Metropolis estimator [40] is applied in order to rate the ability of the tested models to fit the data or to determine the number of clusters M. The integrated likelihood is defined by [40]

$$p(\mathscr{X}|M) = \int \pi(\Theta|\mathscr{X}, M)\, d\Theta = \int p(\mathscr{X}|\Theta, M)\pi(\Theta|M)\, d\Theta \tag{29}$$

where Θ is the vector of parameters, $\pi(\Theta|M)$ is its prior density, and $p(\mathscr{X}|\Theta, M)$ is the likelihood function taking into account that the number of clusters is M.

4 Experimental Results

In our experiments, we performed face recognition using images from the Iris thermal face database[1] which is a subset of the Object Tracking and Classification Beyond the Visible Spectrum (OTCBVS) database. Images are gray-scale infrared of 320 × 240 each and represent different persons under different expressions, poses, and illuminations. Figure 4 shows images from different classes (persons). We applied our method on two datasets. First we used 1,320 images of 15 persons not wearing glasses. Knowing that in IR imaging, thermal radiation cannot transmit through glasses because glasses severely attenuate electromagnetic wave radiation beyond 2.4 μm. For this reason, we decided to investigate if our algorithm will be capable to identify persons with glasses, so we added 880 images of eight persons with glasses.

[1] OTCBVS Benchmark dataset Collection (http://www.cse.ohio-state.edu/otcbvs-bench/).

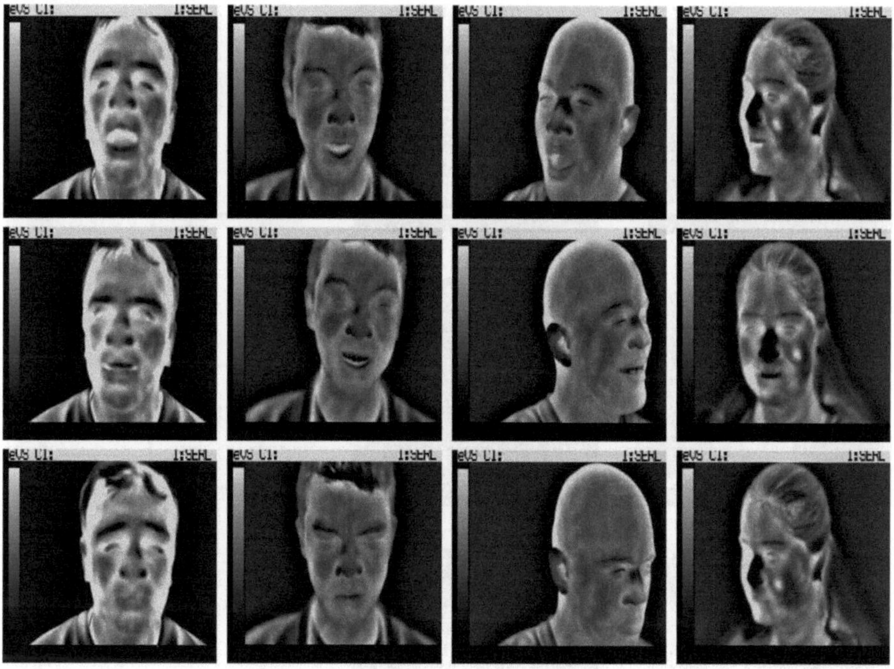

(a) (b) (c) (d)

Fig. 4 Sample images for different classes under different expressions and poses

For both experiments we used 11 images for each person as training set and the rest as testing set. This gave us 165 and 1,155 images for training and testing, respectively, in the first data. The second data set was composed of 253 and 1,947 images for training and testing, respectively.

Face recognition can be considered as an image classification problem where we are trying to classify to which person this image belongs which is based on two main tasks: feature extraction and similarity measurements. Feature extraction allows to generate a set of features known as image signatures capable of representing the visual content of each image in a lower dimensional space. Similarity measurements are used in order to affect each image to the right group. For feature extraction step we have employed both the edge orientation histograms which provide spatial information [17] and the co-occurrence matrices which capture the local spatial relationships between gray levels [34].

Edge orientation histograms are used in order to describe the shape information contained in the images on the basis of its significant edges. We started first by applying Canny edge operator [14], then a histogram of edge directions is used to represent the shape attribute. The main problem of the edge directions histogram is that it is not scale or rotation invariant. In order to solve the problem of scale invariant, we normalized the edge histograms with respect to the number of edge points in

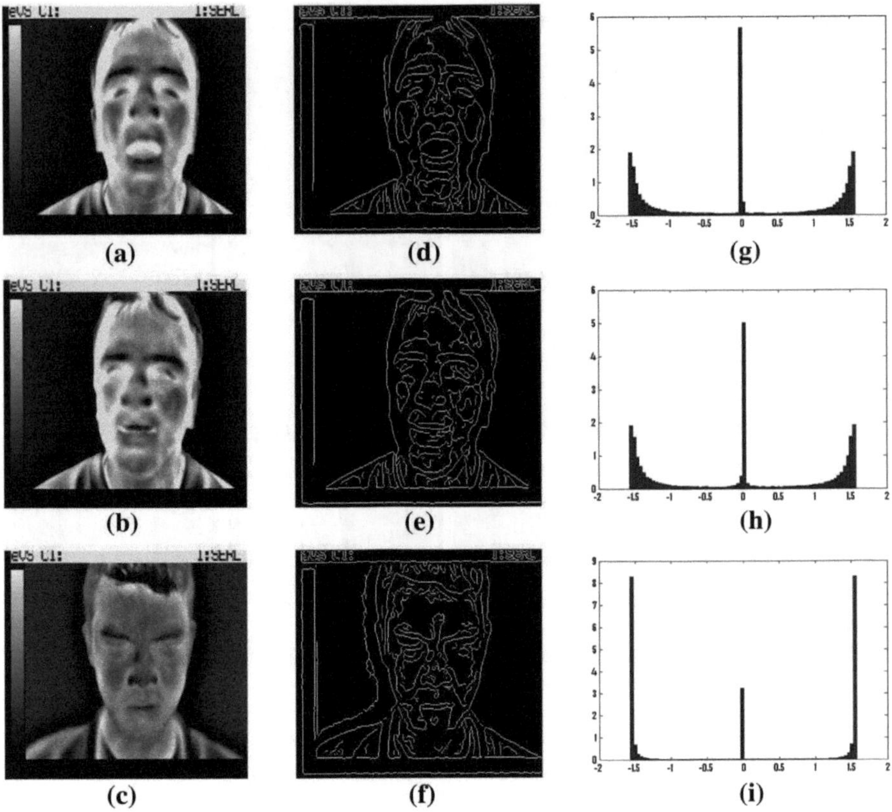

Fig. 5 Three different images of two different persons with their corresponding edge images and corresponding edge histograms, **a–c** show the three images, **d–f** show the corresponding edge images, **g–i** show the corresponding edge histograms

the image. Figure 5 shows three face images, where the first two images are for the same person taken from different poses, and the third image is for another person. It is quite clear that the first two images have very close edge-orientation histograms compared to the third image. Texture is another essential feature widely used in image classification and object recognition. In order to model the texture features we have determined the vector of characteristics for each image, we have computed a set of features derived from the co-occurrence matrices. It is well-known that to obtain good results, several co-occurrence matrices should be computed, each one considering a given neighborhood and direction. In our experiments, we have considered the following four neighborhoods: $(1;0)$, $(1, \pi/4)$, $(1, \pi/2)$, and $(1, 3\pi/4)$, respectively, which generally provide good results [30]. For each of these neighborhoods, we calculated the corresponding co-occurrence, then derived from it the following features: mean, variance, energy, correlation, entropy, contrast, homogeneity, and cluster prominence [30]. Besides, the edge directions were quantized into 72 bins

Table 1 Accuracies of the seven different methods tested

	BGGM (%)	EMGGM (%)	BGMM (%)	EMGMM (%)	PCA (%)	HICA (%)	LICA (%)
First dataset	96.02	94.20	86.67	85.89	95.58	95.32	94.46
Second dataset	95.33	92.40	82.54	82.18	94.35	93.99	92.81

of five each. Using the co-occurrence matrices and the histogram of edge directions each image was represented by a 110-dimensional vector. Our Bayesian approach was then used to model each class in the training set by a finite mixture of generalized Gaussian distributions.

In order to perform the assignments of the images in the test set to the different classes, we have used the following rule: $\mathbf{X} \mapsto \arg\max_k p(\mathbf{X}|\Theta_k)$, where $\mathbf{X}$ is a 110-dimensional vector of features representing an input test image to be assigned to a class k and $p(\mathbf{X}|\Theta_k)$ is a mixture of distributions representing a given class k. In order to validate our Bayesian algorithm (BGGM) we have compared it with the expectation maximization (EM) one (EMGGM). We also compared it to six other methods namely principal component analysis (PCA) with cosine distance, localized independent component analysis (LICA) with cosine distance, holistic ICA (HICA) with cosine distance as implemented by FastICA [1], Bayesian Gaussian mixture models (BGMM) and Gaussian mixture models learned with EM (EMGMM). Table 1 shows the accuracies for the seven different methods. According to this table it is clear that the Bayesian learning outperforms the deterministic one, using EM, when applied to the generalized Gaussian or the Gaussian mixture models. It is clear also the BGGM approach provides the best results which can be explained by its ability to incorporate prior information during classes learning and modeling.

5 Conclusion

In this chapter, we have presented a new algorithm for infrared face recognition based on Bayesian learning of finite generalized Gaussian mixture models. The specific choice of the generalized Gaussian distribution is motivated by its flexibility compared to the Gaussian and by the fact that infrared images statistics are generally non-Gaussian. We have used a Monte Carlo Markov Chain simulation technique based on Gibbs sampling mixed with a Metropolis–Hasting step for parameters estimation. In contrast with the EM algorithm which tries to estimate a single "best" model which is not generally realistic, Bayesian approaches take into account the fact that the data may suggest many "good" models and then consider the average result computed over several models. For the selection of number of clusters we have used the integrated likelihood. The experimental results show the effectiveness of the proposed method. Future work can be devoted to the development of a variational framework for the learning of the generalized Gaussian mixture model and its application to other challenging problems such as fusion of visual and thermal spectra modalities.

References

1. Hyvrinen, A.: The fixed-point algorithm and maximum likelihood estimation for independent component analysis. Neural Process. Lett. **10**(1), 1–5 (1999)
2. Jain, A., Bolle, R., Pankanti, S.: Biometrics: Personal Identification in Networked Society. Kluwer Academic Publishers, Dordrecht (1999)
3. Aiazzi, B., Alpaone, L., Baronti, S.: Estimation based on entropy matching for generalized Gaussian PDF modeling. IEEE Signal Process. Lett. **6**(6), 138–140 (1999)
4. Ben-Arie, N.D.: A volumetric/iconic frequency domain representation for objects with application for pose invariant face recognition. IEEE Trans. Pattern Anal. Mach. Intell. **20**(5), 449–457 (1998)
5. Robert, C.P.: The Bayesian Choice From Decision-Theoretic Foundations to Computational Implementation, 2nd edn. Springer, Berlin (2007)
6. Robert, C.P., Casella, G.: Monte Carlo Statistical Methods, 2nd edn. Springer, Berlin (2004)
7. Scolinsky, D.A., Selinger, A.: A comparative analysis of face recognition performance with visible and thermal infrared imagery. In: Proceedings of the International Conference on Pattern Recognition (ICPR), pp. 217–222 (2002)
8. Scolinsky, D.A., Selinger, A.: Thermal face recognition over time. In: Proceedings of the International Conference on Pattern Recognition (ICPR), pp. 187–190 (2004)
9. Socolinsky, D., Selinger, A., Neuheisel, J.: Face recognition with visible and thermal infrared imagery. Comput. Vis. Image Understanding **91**, 72–114 (2003)
10. Müller, F.: Distribution shape of two-dimensional DCT coefficients of natural images. Electron. Lett. **29**(22), 1935–1936 (1993)
11. Prokoski, F.: History, current status, and future of infrared identification. In: Proceedings of the IEEE Workshop on Computer Vision Beyond the Visible Spectrum: Methods and Applications (CVBVS), pp. 5–14 (2000)
12. Friedrich, G., Yeshurun, Y.: Seeing People in the Dark: Face Recognition in Infrared Images. Springer, Berlin (2003)
13. Pavlidis, I., Symosek, P.: The imaging issue in an automatic face/disguise detection system. In: Proceedings of the IEEE Workshop on Computer Vision Beyond the Visible Spectrum: Methods and Applications (CVBVS), pp. 15–24 (2000)
14. Canny, J.: A computational approach to edge detection. IEEE Trans. Pattern Anal. Mach. Intell. **8**, 679–698 (1986)
15. Miller, J.H., Thomas, J.B.: Detectors for discrete-time signals in non-Gaussian noise. IEEE Trans. Inf. Theory **18**(2), 241–250 (1972)
16. Wilder, J., Phillips, P.J., Jiang, C., Wiener, S.: Comparison of visible and infra-red imagery for face recognition. In: Proceedings of the International Conference on Automatic Face and Gesture Recognition (FG), pp. 182–187 (1996)
17. Jain, A.K., Vailaya, A.: Image retrieval using color and shape. Pattern Recognition **29**(8), 1233–1244 (1996)
18. Ghosh, J.K., Delampady, M., Samanta, T.: An Introduction to Bayesian Analysis Theory and Methods. Springer, Berlin (2006)
19. Kokkinakis, K., Nandi, A.K.: Exponent parameter estimation for generalized Gaussian probability density functions with application to speech modeling. Signal Process. **85**(9), 1852–1858 (2005)

20. Sharifi, K., Leon-Garcia, A.: Estimation of shape parameter for generalized Gaussian distributions in subband decomposition of video. IEEE Trans. Circuits Syst. Video Technol. **5**(1), 52–56 (1995)
21. Trujillo, L., Olague, G., Hammoud, R., Hernandez, B.: Automatic feature localization in thermal images for facial expression recognition. In: Proceedings of the IEEE International Workshop on Object Tracking and Classification Beyond the Visible Spectrum (OTCBVS) (2005)
22. Wolff, L., Socolinsky, D., Eveland, C.: Quantitative measurement of illumination invariance for face recognition using thermal infrared imagery. In: Proceedings of the IEEE Workshop on Computer Vision Beyond the Visible Spectrum: Methods and Applications (CVBVS) (2001)
23. Varanasi, M.K., Aazhang, B.: Parametric generalized Gaussian density estimation. J. Acoust. Soc. Am. **86**(4), 1404–1415 (1989)
24. Do, M.N., Vetterli, M.: Wavelet-based texture retrieval using generalized Gaussian density and Kullback–Leibler distance. IEEE Trans. Image Process. **11**(2), 146–158 (2002)
25. Pi, M.: Improve maximum likelihood estimation for subband GGD parameters. Pattern Recognition Lett. **27**(14), 1710–1713 (2006)
26. Allili, M.S., Bouguila, N., Ziou, D.: Finite generalized Gaussian mixture modeling and applications to image and video foreground segmentation. In: Proceedings of the Canadian Conference on Robot and Vision (CRV), pp. 183–190 (2007)
27. Allili, M.S., Bouguila, N., Ziou, D.: Online video foreground segmentation using generalized Gaussian mixture modeling. In: Proceedings of the IEEE International Conference on Signal Processing and Communications (ICSPC), pp. 959–962 (2007)
28. Allili, M.S., Bouguila, N., Ziou, D.: A robust video foreground segmentation by using generalized Gaussian mixture modeling. In: Proceedings of the Canadian Conference on Robot and Vision (CRV), pp. 503–509 (2007)
29. Allili, M.S., Bouguila, N., Ziou, D.: Finite general Gaussian mixture modeling and application to image and video foreground segmentation. J. Electron. Imaging **17**(1), 1–13 (2008)
30. Unser, M.: Filtering for texture classification: a comparative study. IEEE Trans. Pattern Anal. Mach. Intell. **8**(1), 118–125 (1986)
31. Farvardin, N., Modestino, J.W.: Optimum quantizer performance for a class of non-Gaussian memoryless sources. IEEE Trans. Inf. Theory **30**(3), 485–497 (1984)
32. Arandjelović, O., Hammoud, R., Cipolla, R.: Multi-sensory face biometric fusion (for personal identification). In: Proceedings of the IEEE International Workshop on Object Tracking and Classification Beyond the Visible Spectrum (OTCBVS) (2006)
33. Buddharaju, P., Pavlidis, I., Kakadiaris, I.: Face recognition in the thermal infrared spectrum. In: Proceedings of the IEEE International Workshop on Object Tracking and Classification Beyond the Visible Spectrum (OTCBVS) (2004)
34. Shanmugam, K., Haralick, R.M., Dinstein, I.: Texture features for image classification. IEEE Trans. Syst. Man Cybernet. **3**, 610–621 (1973)
35. Kong, S.G., Heo, J., Abidi, B.R., Paik, J., Abidi, M.: Recent advances in visual and infrared face recognition—a review. Comput. Vis. Image Understanding **97**, 103–135 (2005)
36. Mallat, S.G.: A theory for multiresolution signal decomposition: the wavelet representation. IEEE Trans. Pattern Anal. Mach. Intell. **11**(7), 674–693 (1989)
37. Geman, S., Geman, D.: Stochastic relaxation, Gibbs distributions and the Bayesian restoration of images. IEEE Trans. Pattern Anal. Mach. Intell. **6**(6), 721–741 (1984)
38. Fan, S.-K.S., Lin, Y.: A fast estimation method for the generalized Gaussian mixture distribution on complex images. Comput. Vis. Image Understanding **113**(7), 839–853 (2009)
39. Meignen, S., Meignen, H.: On the modeling of small sample distributions with generalized Gaussian density in a maximum likelihood framework. IEEE Trans. Image Process. **15**(6), 1647–1652 (2006)
40. Lewis, S.M., Raftery, A.E.: Estimating Bayes factors via posterior simulation with the Laplace–Metropolis estimator. J. Am. Stat. Assoc. **90**, 648–655 (1997)

41. Chen, X., Flynn, P.J., Bowyer, K.W.: PCA-based face recognition in infrared imagery: baseline and comparative studies. In: Proceedings of the IEEE International Workshop on Analysis and Modeling of Faces and Gestures (AMFG), pp. 127–134 (2003)
42. Adini, Y., Moses, Y., Ullman, S.: Face recognition: the problem of compensating for changes in illumination direction. IEEE Trans. Pattern Anal. Mach. Intell. **19**(7), 721–732 (1997)
43. Yoshitomi, Y., Miyawaki, N., Tomita, S., Kimura, S.: Facial expression recognition using thermal image processing and neural network. In: Proceedings of the IEEE International Workshop on Robot and Human Communication, pp. 380–385 (1997)
44. Tian, Y.I., Kanade, T., Cohn, J.F.: Recognizing action units for facial expression analysis. IEEE Trans. Pattern Anal. Mach. Intell. **23**(2), 97–115 (2001)
45. Gao, Z., Belzer, B., Villasenor, J.: A comparison of the Z, E_8 and Leech lattices for quantization of low-shape-parameter generalized Gaussian sources. IEEE Signal Processing Lett. **2**(10), 197–199 (1995)

Part II
Multi-Sensor Fusion and Smart Sensors

Fusion of a Camera and a Laser Range Sensor for Vehicle Recognition

Shirmila Mohottala, Shintaro Ono, Masataka Kagesawa and Katsushi Ikeuchi

Abstract Fusing different sensory data into a singular data stream is not a recent idea, but with the diffusion of various simple and compact sensors, multi-sensor fusion has inspired new research initiatives. Sensor fusion improves measurement precision and perception, offering greater benefits than using each sensor individually. In this chapter we present a system that fuses information from a vision sensor and a laser range sensor for detection and classification. Although the laser range sensors are good at localizing objects accurately, vision images contain more useful features to classify the object. By fusing these two sensors, we can obtain 3D information about the target object, together with its textures, with high reliability and robustness to outdoor conditions. To evaluate the performance of the system, it is applied to recognition of on-street parked vehicles from a moving probe vehicle. The evaluation experiments show obviously successful results, with a detection rate of 100% and accuracy over 95% in recognizing four vehicle classes.

Keywords Sensor fusion · Vehicle classification

S. Mohottala · S. Ono · M. Kagesawa (✉) · K. Ikeuchi
Institute of Industrial Science, University of Tokyo, 4-6-1, Komaba Meguro-ku, Tokyo 153-8505, Japan
e-mail: kagesawa@cvl.iis.u-tokyo.ac.jp

S. Mohottala
e-mail: shirmi@cvl.iis.u-tokyo.ac.jp

S. Ono
e-mail: onoshin@cvl.iis.u-tokyo.ac.jp

K. Ikeuchi
e-mail: ki@cvl.iis.u-tokyo.ac.jp

R. Hammoud et al. (eds.), *Machine Vision Beyond Visible Spectrum*, Augmented Vision and Reality, 1, DOI: 10.1007/978-3-642-11568-4_6,

1 Introduction

Vision sensors offer a number of advantages over many other sensors, and they are particularly good at wide and coarse detection. But they have limited robustness in outdoor conditions. Further, dealing with on-board cameras is much more complicated than with stationary cameras, because when the camera is moving, the environment changes significantly from scene to scene, and there is no priori knowledge of the background. On the other hand, recent laser range sensors have become simpler and more compact, and their increasing reliability has attracted more attention in various research applications. They are very efficient in extracting 3D geometric information, but one main drawback of the laser sensors is their lack of capability in retrieving textures of objects.

However, fusion of a vision sensor and a laser range sensor enables the system to obtain 3D information about the scanned objects together with their textures, offering a wide range of other advantages:

- Vision-based algorithms alone are not yet powerful enough to deal with quickly changing or extreme environmental conditions, but when fused with laser range data, robustness and reliability are greatly increased.
- Combined sensory data can be used to enhance perception or to highlight the areas of interest.
- Various parameters can be obtained at once.

Laser and vision sensor fusion has been an active area of research in the field of robotics. In [1, 2] laser range data are used to detect moving objects, and the object is extracted based on the position information. Refs. [3, 4] use laser and stereoscopic vision data for robot localization and motion planning. In [5], a sensor-fused vehicle detection system is proposed that determines the location and the orientation of vehicles using laser range information, and applies a contour-based pattern recognition algorithm. In this chapter we present a system that fuses vision and laser range data for detection and classification of on-street parked vehicles.

On-street parking has been recognized as one of the major causes of congestion in urban road networks in Japan. In order to determine effective strategies to minimize the problem, the road administrators need various statistics regarding on-street parking. Currently there are no automated systems that can scan road situations, so the surveys on road conditions are done manually by human operators. Manual counting is time-consuming and costly. Motivated by this strong requirement, we propose a system that can scan the roadside environment and extract information about on-street parked vehicles automatically. Roadside data is acquired using both a laser range sensor and a vision sensor mounted on a probe vehicle that runs along the road, and then the two sets of scanned data are fused for better results.

In addition, the system will be modified in our future work to extract buildings or road signs from 3D data and to retrieve their textures as well. These capabilities are extremely important for many ITS applications such as 3D navigation systems, 3D urban modeling, and digital map construction.

Fig. 1 System configuration

The rest of this chapter is organized as follows. In Sect. 2, the system configuration and the sensors employed for scanning are explained. Next, starting from Sect. 3, the data processing and recognition algorithms, along with the experimental results of each method, are presented in four sections as follows:

- Segmentation of vehicles from laser range data
- Calibration of laser sensor and camera
- Refinement of segmentation result using both laser and image data
- Classification of vehicles

Finally, we discuss merits and demerits of the whole system in Sect.7, and summarize in the last section.

2 System Configuration

A laser range sensor and a video camera are mounted on a probe vehicle, close to each other facing same direction, so that both sensors can scan the object at the same time (Fig. 1a). The probe vehicle runs in the next lane to the parked vehicles, and performs scans as shown in Fig. 1b.

- *Laser range sensor*. Laser range sensors measure the time of flight or the phase shift of the laser pulse to determine the distance from the scanner to each point on the object from which the laser reflects. This depth information enables the acquisition of the 3-dimensional geometric of an object. Laser scanning is accurate and instantaneous. There are various types of laser sensors and different ways of scanning. Considering the safety of use on public roads, and the robustness required to scan from a moving vehicle, we used a SICK LMS200 laser range sensor. In order to fit our need to scan while progressing and extract the vehicle shape, we set the sensor transversely as shown in Fig. 1, so that the scanning is done vertically.

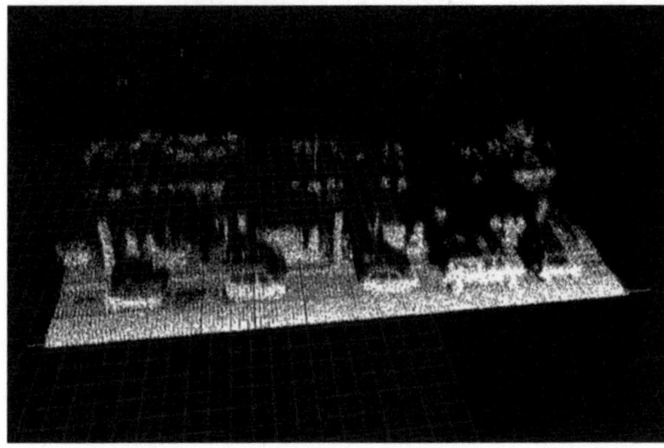

Fig. 2 A depth image

- *Camera*. To acquire vision data, we use a normal video camera with the resolution set to 848×480 and the frame rate to 30 frames per second. A wide view is required because the targeted vehicles need to appear in full within a frame, even when the probe vehicle is close to the parked vehicle.

3 Segmenting Vehicles from Laser Range Data

In this section we present the method for segmenting vehicles from laser range data. The laser range sensor reports laser readings as 3D points in a laser coordinate system, for which the origin is the laser sensor. By fusing with an accurate positioning system like Global Positioning System (GPS) to acquire the position of the sensor, these points can be projected to the World Coordinate System (WCS). Assuming the speed of the probe vehicle is constant and runs straight parallel to parked vehicles, we can line up the scanned lines to display the range data in a depth image as shown in Fig. 2. Here we set the WCS as follows:

- X-axis: the direction the probe vehicle progresses
- Z-axis: the direction vertical to the road surface
- Y-axis: the direction of the outer product of x and z

The laser scan plane is the plane $X = 0$. It would seem that segmenting the vehicle area from the background could be performed simply by cutting off the background on the basis of depth. But this is impossible in some cases. Considering safety, we chose a sensor with a laser that is not very powerful. Hence the reflection of laser from black vehicles is poor, and sometimes only a few points from the body are reflected. A simple detector based on depth will fail to detect those black vehicles. Therefore, we apply a novel method, combining two separate vehicle detectors, which we have presented in our previous work [6].

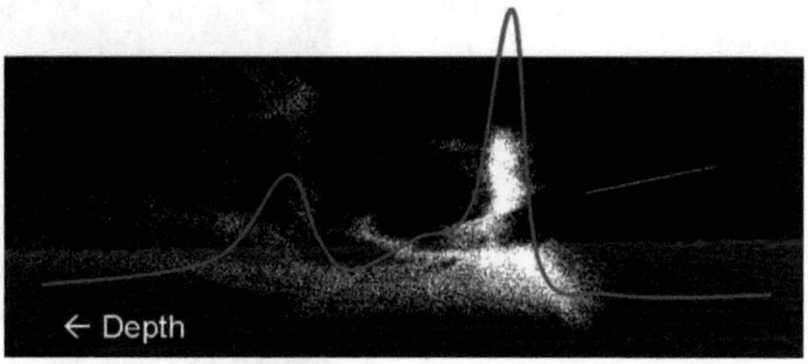

Fig. 3 Histogram of the number of range points when the points are projected to the *Y–Z* plane

3.1 Side Surface-Based Method

When scanned points are projected on to the *Y–Z* plane and the total number of points in the *Y* direction is calculated, we get a histogram as shown in Fig. 3. The side surface of on-street parked vehicles appears in a significant peak. We use this feature to extract vehicles. The side surface of a vehicle *A* is determined as follows:

$$A = \{(x, y, z) \mid z > z_{AB}, y_a < y < y_b\}. \tag{1}$$

By experiment, we set the constants as $Z_{AB} = 0.5\,(\mathrm{m})$, $y_a = 1.0\,(\mathrm{m})$, and $y_b = y_{\mathrm{peak}} + 1.0\,(\mathrm{m})$ and the minimum y that gives a maximum in the histogram is taken as y_{peak}. It is then smoothed with a smoothing filter and only the blocks longer than 2 m are counted as vehicles.

3.2 Silhouette-Based Method

The side surface-based method uses the points of vehicles' side surfaces for detection; hence, the result varies depending on the reflectance of the vehicle body. For example, black vehicles that have a low laser reflectance may not give a significant peak; therefore, they may not be detected. In the silhouette-based method, we focus on the road surface in the depth image. The on-street parked vehicles occlude the road surface, so, as shown in Fig. 4, they make black shadows on the depth image. Despite the reflectance of the vehicle body, these shadows appear because the laser does not go through the vehicles.

The silhouette-based method is as follows: first we extract the road surface by $B = \{(x, y, z) \mid z < z_{AB}\}$ where z_{AB} is set to 0.5 (m). It is then smoothed with a smoothing filter and thresholded to detect vehicle areas. The threshold y_{th} is defined as

$$y_{\mathrm{th}} = \bar{y}_{\mathrm{A}} + 1.5\,(\mathrm{m})$$

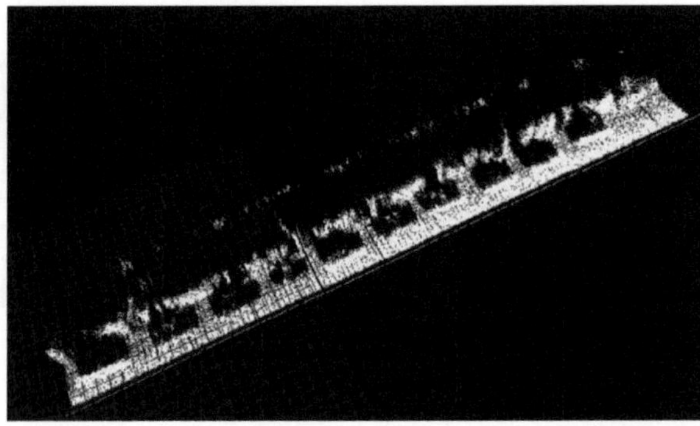

Fig. 4 Depth image of road surface

Fig. 5 Vehicles detected from laser range data

by experiments where $\bar{y}_A$ is the average depth of the segmented region in the side surface-based method.

3.3 Detection Results

The on-street parked vehicles are segmented with two detectors explained above, and the results are combined. In experiments all the vehicles were detected 100% accurately. Figure 5 shows an example of segmented results.

4 Extrinsic Calibration of Camera and Laser Range Sensor

In a sensor-fused system, the process to determine how to combine the data from each sensor is complicated. There are various sensor fusing methods and calibration methods proposed to use the sensors to their fullest potential. Different approaches have been proposed to calibrate cameras and laser range sensors, and most of them use markers or features like edges and corners that can be visible from both sensors [7, 8].

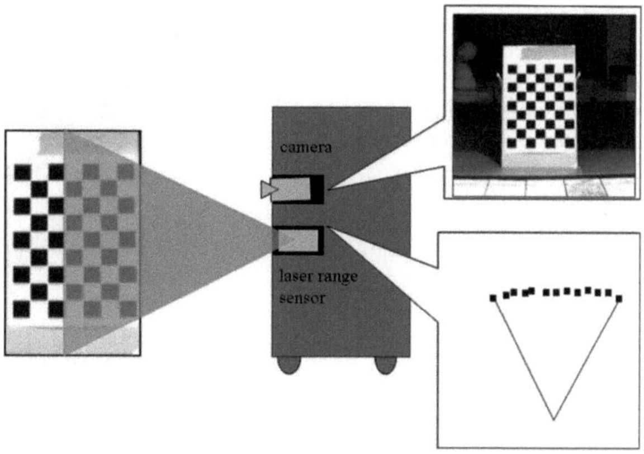

Fig. 6 The setup for calibration of camera and laser range sensor

Some methods use visible laser [9]. Zhang and Pless present a method that makes use of a checkerboard calibration target seen from different unknown orientations [10], and a similar approach is proposed in [11]. Our calibration is based on the method in [10], and only the extrinsic calibration method is presented here, assuming that the intrinsic parameters of the camera are known.

Extrinsic calibration of a camera and a laser range sensor consists of finding the rotation Φ and translation Δ between the camera coordinate system and the laser coordinate system. This can be described by

$$P^l = \Phi P^c + \Delta \tag{2}$$

where P^c is a point in the camera coordinate system that is located at a point P^l in the laser coordinate system. Φ and Δ are calculated as follows.

A planar pattern on a checkerboard setup as shown in Fig. 6 is used for calibration. We refer to this plane as the "calibration plane". The calibration plane should be visible from both camera and laser sensor. Assuming that the calibration plane is the plane $Z = 0$ in the WCS, it can be parameterized in the camera coordinate system by 3-vector N that is vertical to the calibration plane and $\| N \|$ is the distance from the camera to the calibration plane.

From Eq. 2, the camera coordinate P^c can be derived as $P^c = \Phi^{-1}(P^l - \Delta)$ for a given laser point P^l in the laser coordinate system. Since the point P^c is on the calibration plane defined by N, it satisfies $N \cdot P^c = \| N \|^2$.

$$N \cdot \Phi^{-1}(P^l - \Delta) = \| N \|^2 \tag{3}$$

For a measured calibration plane N and laser points P^l, we get a constraint on Φ and Δ.

We extract image data and laser range data in different poses of the checkerboard. Using these data, the camera extrinsic parameters Φ and Δ can refined by minimizing

Fig. 7 The points in one laser-scanned line are projected onto the corresponding image. Laser points are marked in vertical line

$$\sum_i \sum_j \left(\frac{N_i}{\| N_i \|} \cdot (\Phi^{-1}(P^l_{ij} - \Delta)) - \| N_i \| \right)^2 \tag{4}$$

where N_i defines the checkerboard plane in the ith pose. Minimizing is done as a nonlinear optimization problem with the Levenberg–Marquardt algorithm in Matlab.

4.1 Calibration Results

Using the above method we calculate the projection matrix of the laser coordinate system to the image coordinate system. In Sect. 3, the vehicles are segmented from their background in the laser range data. Now we can project these laser range data onto the corresponding images. If the sensor and camera are calibrated correctly, the laser scanned points will be projected onto the vehicle area in images. Figure 7 shows an example where one laser scanned line of a vehicle is projected onto the corresponding image.

If we use an accurate positioning system like GPS along with an INS (Internal Navigation System) to acquire the position of the probe vehicle instantaneously, we can line up all the scanned lines accurately based on the sensor position. This kind of accurate positioning is very useful for applications like 3D town digitalizing. But for our task, a more simple method based on a simple assumption will derive sufficiently accurate results.

Here we assume that the probe vehicle moves straightforward at a constant speed, parallel to on-street parked vehicles. The laser and image data are synchronized manually, by setting the two ends of the data sequences. Next, on the basis of the above assumption, we project all the scanned lines onto the corresponding images, and line them up in constant intervals. Figure 8 presents two examples where the laser-scanned lines are projected onto the vehicles, and the silhouettes of vehicles appear quite accurately. Even though there is a small error, it does not significantly affect our task.

Fig. 8 Some examples of laser-scanned lines projected onto the corresponding images. Scanned points are marked in vertical line

Fig. 9 An example of a black vehicle that is not scanned accurately by laser

5 Refinement of Segmentation

We have already segmented vehicles from laser range data and fused the result with image data. By linking all the maximum points and the minimum points of each line projected onto the images, we can segment the vehicles from images as well. But Fig. 9 shows an example in which this will not work. Because the laser sensor we used is not very powerful, some laser points are not reflected well. This was particularly true for black vehicles, in which the total number of scanned points was less than half the number for white vehicles. Therefore, we employ a vision-based segmentation method as well to refine these segmentation results and enhance the robustness.

5.1 Graph Cut Method

Graph-cut method is a powerful optimizing technique that can achieve robust segmentation even when foreground and background color distributions are not well separated. Greig et al. [12] were the first who proposed the ability to apply the graph-based powerful min-cut/max-flow algorithms from combinatorial optimization to minimize certain important energy functions in vision.

Let $\mathcal{G} = (\mathcal{V}, \mathcal{E})$ be a directed graph with vertices $\mathcal{V}$, the set of elements to be segmented, and edges $e \in \mathcal{E}$ corresponding to pairs of neighboring vertices. Each edge e is assigned a nonnegative weight (cost) w_e. In the case of image segmentation, the elements in V are pixels, and the weight of an edge is some measure of the dissimilarity between the pixels connected by that edge. There are two special terminals: an "object" and a "background". The source is connected by edges to all nodes identified as object seeds, and the sink is connected to all background seeds. A cut is a subset of edges, which has a cost that is defined as the sum of the weights of the edges that it severs

$$|C| = \sum_{e \in C} w_e$$

The segmentation boundary between the object and the background is drawn by finding the minimum costs cut on the graph.

5.2 Results of Segmentation Using Laser Range Data and Graph Cut Method

We apply the graph cut method to refine the segmentation results obtained earlier by only projecting segmented laser data. The graph cut method requires a user-initiated indication of background/foreground areas. Foreground is denoted by the earlier segmentation result, based on the points segmented from laser data and projected onto the images. Note that the points closer to the outline on the top and the bottom of a scanned line are ignored, considering the calibration error. Then two lines are drawn on the top and the bottom of the image in red, indicating the background area. These two lines are fixed for all the images and drawn horizontally at $y = 50$ and $y = height - 50$. Figure 10 shows some examples of segmentation results. The background is painted over in purple. Despite the very few number of points extracted from the black vehicle, this method enabled accurate segmentation of the vehicle area.

Fig. 10 Some examples of vehicle segmentation using laser range data and graph cut method

6 Classification of On-Street Parked Vehicles

This section describes how the system recognizes the classes (categories) of the vehicles segmented above. The classification algorithm is based on our previous work [13], which classified vehicles from an overhead surveillance camera. Compared to the top view, side views of vehicles give more distinguishing features for classification, so the task seems to be easy. But one main drawback that interferes with the classification process is the reflectance from the vehicle body surface. Important features of some vehicles are barely visible due to reflection of sunlight, while some vehicles appear with very strong noise due to reflection of the surroundings. Moreover, in some vehicles, accessories such as the shape of the doors, lights, mirrors or wheel caps appear with unique features, changing the appearance of the whole vehicle. To deal with these characteristics, we apply some modifications to the previous method, and evaluate the system for recognition of four very similar vehicle classes: sedan, minivan, hatchback, and wagon.

6.1 Classification Method

The classification algorithm is based on the Binary Features method [14], which can be introduced as a modified step of eigen window technique [15]. In this method, Principal Component Analysis (PCA) that uses real valued eigenvectors is replaced with the vector quantization method, to avoid floating point computations and reduce the required computation time and memory. Further, binary edge images are used instead of gray level images, increasing robustness to illumination changes. First, the binary edge images are calculated from the original image, using a Laplacian and Gaussian filter. Then the binary features are selected as described below.

Let $B(e_i(x, y); x_0, y_0; b)$ be a window of size $(2b + 1) \times (2b + 1)$ pixels around (x_0, y_0) in a binary edge image $e_i(x, y)$. Each window is rated as follows,

$$r_i(x, y) = \min_{\substack{-d \le d_x \le d \\ -d \le d_y \le d}} \left\{ D_H \left[\mathbf{\Omega}\{e_i(x', y'); x, y, b\}, \right. \right.$$
$$\left. \left. \mathbf{\Omega}\{e_i(x', y'); x + d_x, y + d_y, b\} \right] \right\} \quad (d_x, d_y) \neq (0, 0) \tag{5}$$

where $D_H(\bar{A}, \bar{B})$ is the Hamming distance between two binary vectors $\bar{A}$ and $\bar{B}$. The window will rate highly if it is dissimilar to its surrounding. Highly rated windows are taken as features, and then compressed to code features using standard Lloyd's algorithm, based on vector quantization. Code features are computed based on the nearest neighbor clusters of training features. For each coded feature, a record of the group of features it represents and the locations of these features on training images is kept.

To detect an object in an input image, first, the input image is processed the same way as the training images to get a binary edge image. The binary edge input image is then encoded, or in other words, the nearest code corresponding to each pixel $c(x, y)$ is searched using

$$c(x, y) = \arg \min_{a \in [1, 2, \ldots, n_c]} \left\{ D_H \left[\bar{\Omega}\{e(x', y'); x, y, b\}, \bar{F}_a \right] \right\} \tag{6}$$

where $\bar{F}_a$ denotes the codes computed using Lloyd's algorithm. The encoded input image is then matched with all the training images to find the nearest object through a voting process.

6.1.1 Voting Process

We prepare a voting space $V(T)$ for each training image T. Let $w(J, X_i, Y_i)$ be a feature at (X_i, Y_i) on the input image J. First we find the nearest code feature for w, and the corresponding features $w_n(T_k, x_i', y_i')$ represented by that code. w_n is a feature at (x_i', y_i') on the training image T_k. Once the best matched feature is found, we put a vote onto the point $(X_i - x_i', Y_i - y_i')$ of the vote space $V(T_k)$ (Fig. 11). Note that one vote will go to the corresponding points of each feature represented by that code.

If the object at (X, Y) in the input image is similar to the object in the training image T, a peak will appear on $V(T)$ at the same location (X, Y). The object can be detected by thresholding the votes.

As our input images are not in a high resolution, the votes may not stack up exactly on one point, but they will concentrate in one area. Therefore, we apply a Gaussian filter to the voting space, improving the robustness of the object scale (Figs. 12 and 13).

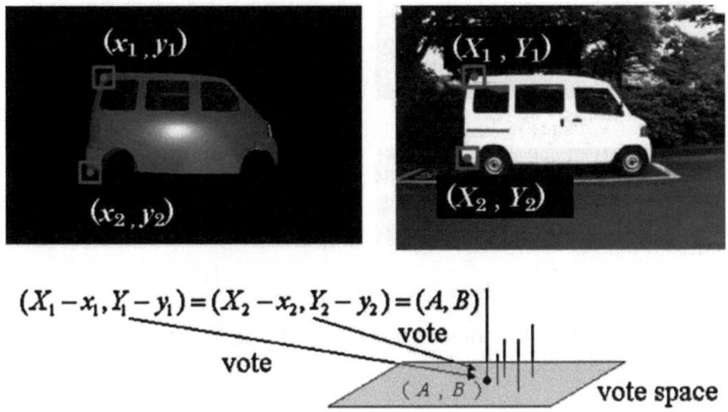

Fig. 11 Voting process

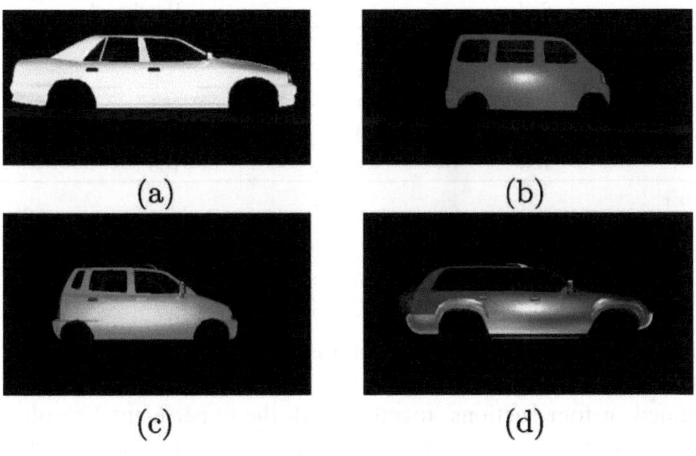

Fig. 12 Some of the CG training images used for classification of on-street parked vehicles: **a** sedan; **b** minivan; **c** hatchback; **d** wagon

6.2 Classification Results

The system is evaluated through experiments using a set of images with 37 sedans, 28 minivans, 31 hatchbacks, and 43 wagons, a total of 139 images. Classification results are denoted in a confusion matrix in Table 1. Some examples of successfully classified images are presented in Fig. 14. The input edge image is shown in gray and the training feature image with the highest match is overlapped and drawn in white. The total classification ratio of the system is 96%. Figure 15 shows an example of the robustness of this method to partial occlusion.

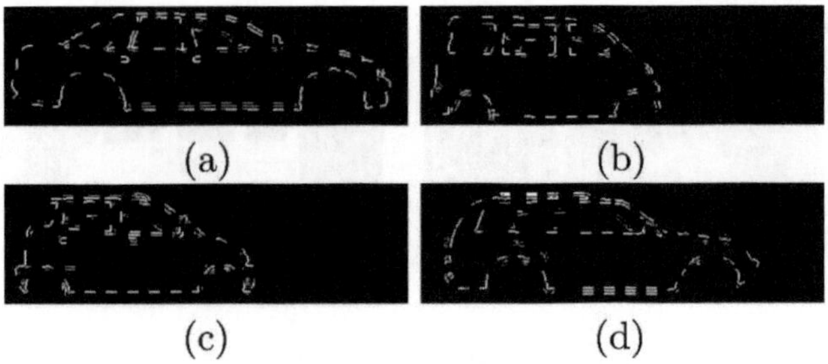

Fig. 13 Features extracted from CG training images for classification of on-street parked vehicles: **a** sedan; **b** minivan; **c** hatchback; **d** wagon

Table 1 Classification results of on-street parked vehicles when a few recent minivan models are eliminated from the input image set

Classified as	Real class			
	Sedan	Minivan	Hatchback	Wagon
Sedan	37	0	0	1
Minivan	0	25	0	0
Hatchback	0	3	31	1
Wagon	0	0	0	41
Accuracy (%)	100	89	100	95

Accuracy: 96%

7 Discussion

We presented a recognition system that fuses a laser range sensor and a camera to detect and classify on-street parked vehicles. The processing algorithms were explained in four sections, together with the experimental results. Here we discuss the characteristics and merits of each algorithm, following with issues and future work.

The method to segment the vehicles from laser range data worked very well, detecting the vehicles 100% accurately. The calibration of the two sensors was quite successful, as we can see in experimental results. When projecting laser data onto the images, the sensor position is calculated assuming the probe vehicle runs at a constant speed. But in reality, the speed of the probe vehicle at each point is not constant, and this led to a small error in synchronizing, even though it did not influence the performances of the system. In our future work, we plan to fuse a positioning system to acquire the real position of the probe vehicle, so that we will be able to derive more precise results. The graph cut method modified by fusing laser-segmented results to initialize the foreground/ background area could segment the vehicle area from its background accurately. Even the black vehicles where the laser range data are not reflected well were segmented correctly. We believe that our method is robust enough to extract road signs or textures of buildings automatically, which we hope to achieve in our future work.

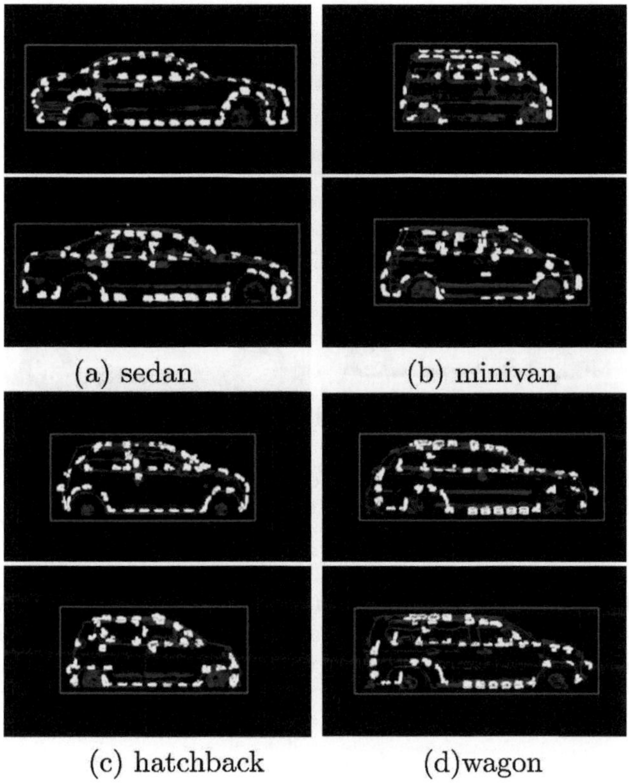

Fig. 14 Some examples that are classified successfully. Input vehicle is shown in *gray* and the training vehicle best matched is indicated in *white*: **a** sedan; **b** minivan; **c** hatchback; **d** wagon

Fig. 15 A successful classification of a partially occluded vehicle

Classification of on-street parked vehicles showed very good results with an accuracy of 96%. Sedans and hatchbacks achieved a significant classification rate with no failures, while wagons showed a high accuracy of 95%. The classification rate was poor only in minivans. A box-shaped body, larger than sedans or wagons in height and with three rows of seats were some features of the class "Minivan" we set. The real class of each vehicle model was determined according to automobile manufacturers. But the shape of the recent minivans have changed considerably; they are not box-shaped any more. Their front ends look similar to wagons, and height

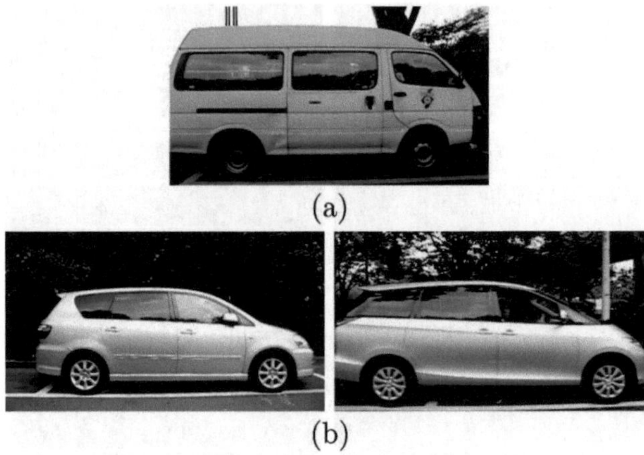

Fig. 16 How the model of recent minivans differs from a conventional model. **a** a conventional minivan with box-shaped body; **b** recent models of minivans

does not differ much from wagons. Figure 16 shows an example of a conventional minivan and some recent models of minivans that appeared in our experimental data. In future work, we hope to reconsider the basis of these classes to fit today's vehicle models.

8 Conclusion

We presented a recognition system that fuses a laser range sensor and a camera for scanning, and we used the resulting types of information to detect and recognize the target objects. The system was evaluated on its success in detecting and classifying on-street parked vehicles, and it achieved a total classification accuracy of 96%. The sensor-fused system we proposed here can model the 3D geometric of objects and segment their texture from the background automatically, so that the texture can be mapped onto the 3D model of the object. In our future work, we hope to fuse this system with a positioning system like GPS and INS and apply this technique for digitalizing and 3D modeling of urban environments, as a part of our ongoing project.

References

1. Blanco, J., Burgard, W., Sanz, R., Fernandez, J.L.: Fast face detection for mobile robots by integrating laser range data with vision. In: Proceedings of the IEEE International Conference on Robotics and Automation, pp. 625–631 (2003)

2. Song, X., Cui, J., Zhao, H., Zha, H.: Bayesian fusion of laser and vision for multiple people detection and tracking. In: Proceedings of the SICE Annual Conference on 2008 (SICE08), pp. 3014–3019 (2008)
3. Baltzkis, H., Argyros, A., Trahanias, P.: Fusion of laser and visual data for robot motion planning and collision avoidance. Mach. Vis. Appl. **15**, 92–100 (2003)
4. Pagnottelli, S., Taraglio, S., Valigi, P., Zanela, A.: Visual and laser sensory data fusion for outdoor robot localisation and navigation. In: Proceedings of the 12th International Conference on Advanced Robotics, 18–20 July 2005, pp. 171–177 (2005)
5. Wender, S., Clemen, S., Kaempchen, N., Dietmayer, K.C.J.: Vehicle detection with three dimensional object models. In: IEEE International Conference on Multisensor Fusion and Integration for Intelligent Systems, Heidelberg, Germany, September (2006)
6. Ono, S., Kagesawa, M., Ikeuchi, K.: Recognizing vehicles in a panoramic range image. In: Meeting on Image Recognition and Understanding (MIRU), pp. 183–188 (2002)
7. Mei, C., Rives, P.: Calibration between a central catadioptric camera and a laser range finder for robotic applications. In: Proceedings of ICRA06, Orlando, May (2006)
8. Wasielewski, S., Strauss, O.: Calibration of a multi-sensor system laser rangefinder/camera. In: Proceedings of the Intelligent Vehicles '95 Symposium, pp. 472–477 (1995)
9. Jokinen, O.: Self-calibration of a light striping system by matching multiple 3-d profile maps. In: Proceedings of the 2nd International Conference on 3D Digital Imaging and Modeling, pp. 180–190 (1999)
10. Zhang, Q., Pless, R.: Extrinsic calibration of a camera and laser range finder (improves camera calibration). In: Intelligent Robots and Systems (IROS), (2004)
11. Unnikrishnan, R., Hebert, M.: Fast extrinsic calibration of a laser rangefinder to a camera. Technical Report, CMU-RI-TR-05-09, Robotics Institute, Carnegie Mellon University, July (2005)
12. Greig, D., Porteous, B., Seheult, A.: Exact maximum a posteriori estimation for binary images. J. R. Stat. Soc. **51**(2), 271–279 (1989)
13. Yoshida, T., Mohottala, S., Kagesawa, M., Tomonaka, T., Ikeuchi, K.: Vehicle classification system with local-feature based algorithm using cg model images. IEICE Trans. Inf. Syst. **E85D**(11), 1745–1752 (2002)
14. Krumm, J.: Object detection with vector quantized binary features. In: Proceedings of Computer Vision and Pattern Recognition, San Juan, Puerto Rico, pp. 179–185 (1997)
15. Ohba, K., Ikeuchi, K.: Detectability, uniqueness, and reliability of eigen windows for stable verification of partially occluded objects. IEEE Trans. Pattern Anal. Mach. Intell. **19**(9), 1043–1048 (1997)

A System Approach to Adaptive Multimodal Sensor Designs

Tao Wang, Zhigang Zhu, Robert S. Krzaczek and Harvey E. Rhody

Abstract We propose a system approach to adaptive multimodal sensor designs. This approach is based on the integration of tools for the physics-based simulation of complex scenes and targets, sensor modeling, and multimodal data exploitation. The goal is to reduce development time and system cost while achieving optimal results through an iterative process that incorporates simulation, sensing, processing and evaluation. A Data Process Management Architecture (DPMA) is designed, which is a software system that provides a team development environment and a structured operational platform for systems that require many interrelated and coordinated steps. As a case study, we use an effective peripheral–fovea design as an example. This design is inspired by the biological vision systems for achieving real-time target detection and recognition with a hyperspectral/range fovea and panoramic peripheral view. This design will be simulated and evaluated by realistic scene and target simulations, and the related data exploitation algorithms will be discussed.

Keywords Hyperspectral imaging · Panoramic vision · System architecture · Multimodal sensing · Target tracking

1 Introduction

Recently, a great deal of effort has been put into adaptive and tunable multi-spectral or hyper-spectral sensor designs with goals to address the challenging problems of detecting, tracking and identifying targets in highly cluttered, dynamic scenes.

T. Wang (✉) · Z. Zhu
Department of Computer Science, City College of New York, NewYork, NY 10031, USA
e-mail: flyingwave001@hotmail.com

R. S. Krzaczek · H. E. Rhody
Chester F. Carlson Center for Imaging Science, Rochester Institute of Technology, Rochester, NY 14623-5604, USA

R. Hammoud et al. (eds.), *Machine Vision Beyond Visible Spectrum*, Augmented Vision and Reality, 1, DOI: 10.1007/978-3-642-11568-4_7,

Representative large programs include: the DARPA's Adaptive Focal Plane Array (AFPA) Program [10, 12], ARL's Advanced Sensor CTA [7], and NSF's Center for Mid-Infrared Technologies for Health and the Environment [13]. Whereas these efforts mainly focus on semiconductor materials, photonics, and hardware designs, and have created or will soon create novel adaptive multimodal sensors, the sensors that have been designed and are being designed are not up to the expectations for real-world applications. Another piece of novel sensor design will not by itself revolutionarily change this situation. Our work looks into using a system approach to utilize advanced multimodal data exploitation and information sciences for innovative multimodal sensor designs to satisfy the requirements of real-world applications in security, surveillance and inspection.

1.1 What is Needed?

Modern forward-looking infrared (FLIR) imaging sensors can achieve high detection and low false alarm rates through the exploitation of the very high spatial resolution available on current generation of large format plane arrays (FPAs). However, today's HSI systems are limited to scanning mode detection, and are usually large, complex, power hungry and slow. The ability to perform HSI in a staring mode is critical to real-time targeting mission. This brings up conflicting requirements for real-time, large area search with the ability to detect and identify difficult and hidden targets using hyperspectral information, while staying within the processing time and size available suited to small platforms. DARPA's Adaptive Focal Plane Array (AFPA) Program is to develop an electro-optical imaging sensor that benefits from both hyperspectral and FLIR, while avoiding the large mass of hyperspectral and the poor target-to-background signal differential. The AFPA designs allow a pixel-by-pixel wavelength selection in hyper-spectral imaging. With continuous spectral tuning, users may re-program the AFPA based upon the characteristics of the target and background. However, this requires the ability to decide which wavelengths to select. As researchers have not identified any unique band of interest, making the sensor spectral tunable is not sufficient. It may be that measurements should be made at collection time that enables the bands to be selected and the associated algorithms tuned correspondingly. In addition, the limited field of view (FOV) of conventional sensor design does not satisfy the requirements of large area search. With this in mind, instead of starting with what technologies can provide, we start with a single bigger question: what do users really need? Under this new paradigm, the three major sub-questions that we ask in a "performance-driven" context are as the following.

1. *System evaluation*: Given a real-world task, how could we rapidly prototype, optimally utilize and evaluate a multimodal sensor, using a general framework and a set of modeling tools that can perform a thorough and close-loop evaluation of the sensor design?

2. *Sensor description*: Given a real-world task, what are the optimal sensing configurations, subsets of data and data representation that are most decision-relevant to provide guidelines for adaptive multimodal sensor designs?
3. *Data exploitation*: Given a real-world task, what advanced data processing and exploitation are needed to support intelligent data collection?

Based on these we propose a system approach for adaptive sensor designs that is possible to reduce development time and system cost while achieving better results through an iterative process. With this approach, it is possible to reduce development time and system cost while achieving better results through an iterative process that incorporates user requirements, data and sensor simulation, data exploitation, system evaluation and refinement.

1.2 A System Approach

Conventionally, the development of a multimodal sensor system requires that many components be selected and integrated in a manner that fits a task and maximizes performance. Such system includes a variety of design tradeoffs that would be difficult and expensive to determine by building physical prototypes. It is inflexible because of the difficulty in changing early design decisions when that would imply more investigations and trade studies. Furthermore, it is difficult to include the end users in the process and to thoroughly evaluate the sensor performance.

The need of a generic system design can reduce the development time and cost by modeling the components and simulating their response using synthetically generated data. This is implemented through scene and sensor simulation tools to model and simulate the background and target phenomenology and sensor characteristics, and place them in a realistic operational geometry. The proposed framework is based on the Digital Imaging and Remote Sensing Image Generation (DIRSIG) [4] tools for characterizing targets, environments and multimodal sensors. Through a realistic scene simulation and sensor modeling process, ground truth data are available for evaluating the designed sensors and related vision algorithms. The simulation tools also allow us to more effectively refine our sensor designs. A Data Process Management Architecture (DPMA) is designed, which is a software system that provides a team development environment and a structured operational platform for systems that require many interrelated and coordinated steps. As a case study, we use a peripheral–fovea design as an example to show how the evaluation and refinement can be done within a system context. This design is inspired by the biological vision systems for achieving real-time imaging with a hyperspectral/range fovea and panoramic peripheral view. Issues of sensor designs, peripheral background modeling, and target signature acquisition will be addressed. This design and the related data exploitation algorithms will be simulated and evaluated in our general data simulation framework.

This chapter is organized as the following. Section 2 illustrates the system design framework. Section 3 shows the design of the bio-inspired adaptive multi-modal sensor platform—the dual panoramic scanners with hyperspectral/range fovea (DPSHRF) for the task of tracking moving targets in real time. Section 4 describes the simulation environment for implementing our system approach, and the parameter configuration of the sensor platform. Section 5 presents the image exploitation algorithms for detecting and tracking moving targets, and the spectral classification method in recognizing moving objects. Conclusions and discussions will be provided in Sect. 6.

2 System Architecture

The Data Process Management Architecture (DPMA) is a software system that has been under development in an evolutionary manner for the last several years to support data collection, data management and analysis tasks. An early design system was called the Data Cycle System (DCS) for NASA's Stratospheric Observatory For Infra-red Astronomy (SOFIA) [1]. The DCS provides the primary interface of this observatory to the science community and supports proposing, observation planning and collection, data analysis, archiving and dissemination. The RIT Laboratory for Imaging Algorithms and Systems (LIAS) is the lead in the DCS development under contract to the Universities Space Research Association (USRA) and works with team members from UCLA, University of Chicago, NASA ARC and NASA GSFC. A second system was developed and is in operational use to support real-time instrument control, data processing and air-to-ground communications as a part of the Wildfire Airborne Sensor Program (WASP) project. The goal of WASP is to provide a prototype system for the US Forest Service to use in wildfire management. The real-time component for WASP, called the Airborne Data Processor (ADP), was constructed using the knowledge we had gained in doing the DCS project, but it is significantly different. Its real-time processing supports geographic referencing and orthographic projection onto standard maps (e.g., WGS-84), mosaic generation, and detection of events and targets of interest.

The DPMA is a design that is based on the experience with both the SOFIA DCS and WASP ADP as well as other activities related to distributed processing, archiving, computing and collaborative decision support. It provides: (1) an adaptable workflow system, capable of managing many simultaneous processing tasks on large collections of data; (2) a set of key abstractions that allow it to be agnostic regarding both data formats as well as processing tasks.

While analyzing the requirements of a system supporting the long term archival and workflow requirements of sensing and image processing systems, we identified three key abstractions. These are the core elements of the DPMA, and are the basis on which we can offer a system that is flexible in its support of algorithms, scalable in its workload, and adaptive to future growth and usage.

The first element of our architecture is the management of the data itself. Starting with an archive supporting a wide variety of data types (e.g., images, vectors, and shape files), new data types can be added to the system by writing additional front ends to this archive as needed. Data stored in this archive can be available for future processing or exchange with other image processing professionals, and mined in the future to generate indexes as new features become relevant. Most importantly, data in this archive can be grouped together into collections to be processed by various agents and operators; any instance of some particular data may be named in multiple collections, supporting multiple and simultaneous assignments of work in the DPMA. Finally, new data and new versions of existing data are accepted by the archive, but the data it replaces is not lost; in this way, we support historical accuracy and analysis, as well as quality control and evaluation of competing algorithms over time.

The second element of our architecture is the specification of processing agents. Each step in an image processing chain, be it an implementation of an automated algorithm, an interactive tool driven by an imaging expert, or even a simple quality assessment by reviewers, is embodied as a processing agent. We provide a support layer for these agents that provide a mechanism for delivering the materials in a work assignment from the archive to the agent, as well as a similar mechanism for storing all new data products yielded by an agent back in the archive, making them available for another agent. This support of processing agents allows the DPMA to evolve image processing and analysis tasks from those that may be performed by a single operator at an image processing workstation to clusters or specialized hardware implementing developed and groomed algorithms in an automated and unattended fashion.

The third element of our architecture is the definition of the workflows themselves. A workflow will be a directed graph whose nodes are the aforementioned processing agents. In the archive, a work assignment is associated with a workflow, moving through the nodes of its graph to reflect its current processing state. Because an agent implementing a processing step can just as easily be a quality assessment, or "gate", as it can be an image processing or decision algorithm, it is easy to instrument workflows with progress reviews, data evaluations, and so on. Similarly, just by manipulating the state of a work assignment (that is, by moving an assignment to a different node in a workflow graph), it is trivial to repeat previous steps in the workflow with different algorithm parameters or operator instructions until a reviewer is satisfied that the assignment can proceed to the next step in the workflow.

These three resources: bundles of data as a work assignment, intelligent distributed agents, and processing workflows, are orthogonal to each other. They all reference each other, but they are defined independently and separately. Each contains entities that name or reflect entities of the other two elements. Taken separately, each part of the system can be grown independently over time with improvements to existing entities or entirely new entities; this growth does not affect entities elsewhere in the system, and dramatically reduces the typical overall risk of system upgrades. Taken together, we have a system that can be easily adapted to new types of data (archive), new processing steps (agents), or new approaches to solving a problem (workflows).

To further demonstrate our system architecture for managing data process we will first describe a complex sensor design that effectively uses a small hyperspectral fovea to gather only important data information over a large area.

3 A Bio-Inspired Sensor Design

To break the dilemma between FOV and spatial/spectral resolution for applications such as wide-area surveillance, we investigate a bio-inspired data collection strategy, which can achieve real-time imaging with a hyperspectral/range fovea and panoramic peripheral view. This is an extension of the functions of human eyes that have high-resolution color vision in the fovea and black-white, low-resolution target detection in the wide field-of-view peripheral vision. The extension and other aspects of our system are also inspired by other biological sensing systems [15]. For instance, certain marine crustaceans (e.g., shrimp) use hyperspectral vision in a specialized way. In our system, the hyperspectral vision is only for the foveated component. As another example, each of the two eyes of a chameleon searches 360° FOV independently. This inspires us to design two separate panoramic peripheral vision components. Some species (such as bats and dolphins) have excellent range sensing capabilities. We add range sensing in our simulated fovea component for enhance and correct the hyperspectral measurements.

The data volumes in consideration have two spatial dimensions (*X* and *Y*), a spectral dimension (*S*, from a few to several hundred), and a time dimension (*T*). This four dimensional (4D) image in *X–Y–S–T* may be augmented by a 2D range image (in the XY space). Ideally, a sensor should have 360° full spherical coverage, with high spatial and temporal resolution, and at each pixel have full range of spectral and range information. However, this type of sensor is difficult to implement because of the enormous amount of data that must be captured and transmitted, most of which will eventually be discarded. Therefore, particularly for real-time applications, every collection must face fundamental trade-offs such as spatial resolution versus spectral resolution, collection rate versus SNR, field-of-view versus coverage, to name a few examples.

Understanding the trade-offs and using algorithms that can be adapted to changing requirements can improve performance by enabling the collection to be done with maximum effectiveness for the current task. In our design, the fovea is enhanced by HSI and range information, and the peripheral vision is extended to panoramic FOV and has adaptive spectral response rather than just black–white.

Our proposed sensor platform, the dual-panoramic scanners with a hyperspectral/range fovea (DPSHRF) (Fig. 1), consists of a dual-panoramic (omnidirectional) peripheral vision and a narrow FOV hyperspectral fovea with a range finder. This intelligent sensor works as the follows: In the first step, two panchromatic images with 360° FOV are generated by rotating two line scanners around a common rotating axis, pointing apart to two slightly different directions. The angle difference between the two scanners can be adjusted for detecting and tracking moving

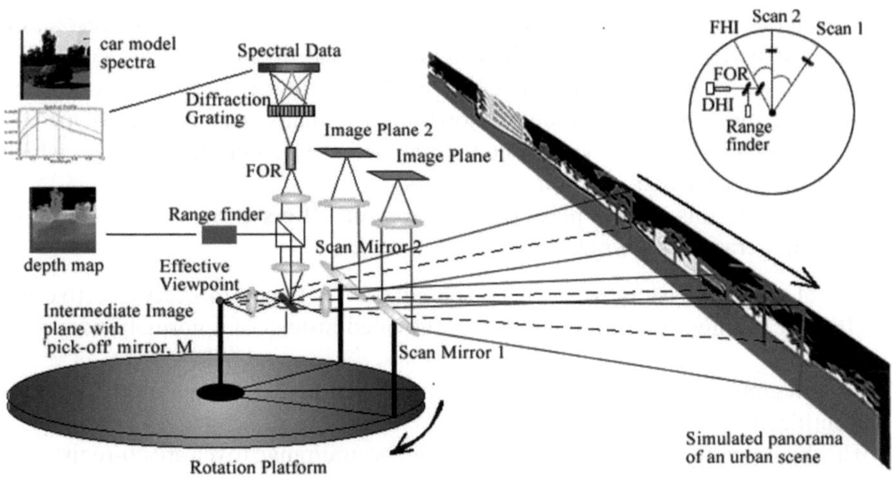

Fig. 1 The design concept of the DPSHRF. The *dash lines* indicate the single viewpoint of both the foveal hyperspectral imager and the two line scanners

targets with different velocities and distances. An initial angle is used at the beginning. Then the detecting results from the two scans can determine what the new angle difference should be—either decreased if a target is moving too fast, or increased if the target is moving too slow. There are two advantages of using line scanners that will be further amplified. First, a line scanner can have a full 360° horizontal FOV. Second, resulted images are inherently registered.

Moving targets can then be easily and quickly determined by the differences of the two panoramic images generated from two scanners. The next position and the time of a moving target can be estimated from the difference of two regions of interest (ROIs) that include the target. In real-time processing, the comparison is started whenever the second scan reaches the position of the first scan, therefore, only a small portion of panoramic images is used before full-view panoramas are generated. The detail of the target detection processing algorithm will be discussed in Sect. 5.

Then, we can turn the hyperspectral/range fovea with a specific focal length calculated based on the size of the object, and to the predicted region that includes the moving target. Thus, hyperspectral/range data is recorded more efficiently for only the ROIs that include possible moving targets. The two line scanners and the hyperspectral/range imager are aligned so that they all share a single effective viewpoint. The spectral data can be efficiently recorded with a foveal hyperspectral imager (FHI) [6] which maps a 2D spatial image into a spatial 1D image. This is implemented by using a micro mirror as a fovea that intercepts the light onto a beam splitter for generating co-registered range-hyperspectral images using a ranger finder and the FHI. The FHI consists of a fiber optical reformatter (FOR) [5] forms a 1D array onto a dispersive hyperspectral imager (DHI) [11] which produces a 2D hyperspectral data array with one dimension as spatial and the other as spectral. The spatial resolution of the FOR is determined by the diameters of optical fibers which are controlled during the optical design process. The blurring effect from cross-coupling of optical fibers is not significant magnitude as shown in [9]. Finally, a co-registered spatial–spectral/range image is produced by combining with the panchromatic images which are generated by the dual-panoramic scanners.

In summary, this sensor platform improves or differs from previous designs [6, 7, 9] in literature in four aspects:

1. A dual scanning system is designed to obtain moving targets in a very effective and efficient manner. A panoramic view is provided instead of a normal wide-angle view.
2. An integration of range and hyperspectral fovea component is used for target identification.
3. The dual-panoramic scanners and the hyperspectral/range fovea are co-registered.
4. Active control of the hyperspectral sensor is added to facilitate signature acquisition of targets of various locations that can only be determined in real-time.

4 Scene Simulation and Sensor Modeling

The sensor design concept is tested though the simulation tool DIRSIG. Various broad-band, multi-spectral and hyperspectral imagery are generated through the integration of a suite of first principles based radiation propagation sub-models [16]. Before performing scene simulation and sensor modeling, we need to set up different scenarios and configure the sensor parameters. One of the complex scenarios we constructed including four cars having exactly the same shape and three different paints moving to different directions with various speeds (Fig. 2). All four cars will pass through the cross section at the bottom corner of the main building in the scene at a certain time. Various behaviors of the moving vehicles such as simple moving, overtaking, passing through, and etc., are monitored by our sensor platform which is placed in front of the main building. The scan speed of each line scanner can be set from 60 to 100 Hz selectable, thus one entire 360° scan take from 6.0 down to 3.6 s. This time constraint is not a problem for real-time target detection since detection and scanning are continuous and simultaneous. The number of pixels per line in the vertical direction is set to 512 to match the horizontal scanning resolution. Few selected spectral bands are captured by dual line scanning. The focal length is fixed at 35 mm for both line scanners, and the angle between the pointing directions of the two scanners is 10° so that the time the second scan reaches the position of the first scan is only about 0.1 s. In theory, the time difference between two scans should be much less than 1 s to avoid a lot of uncertainty of action changes in moving vehicles. Two scanners are used so that (1) the more accurate direction and the focal length of the hyperspectral fovea can be estimated; and (2) moving target detection can still be performed when background subtraction using a single scanner fails due to cluttered

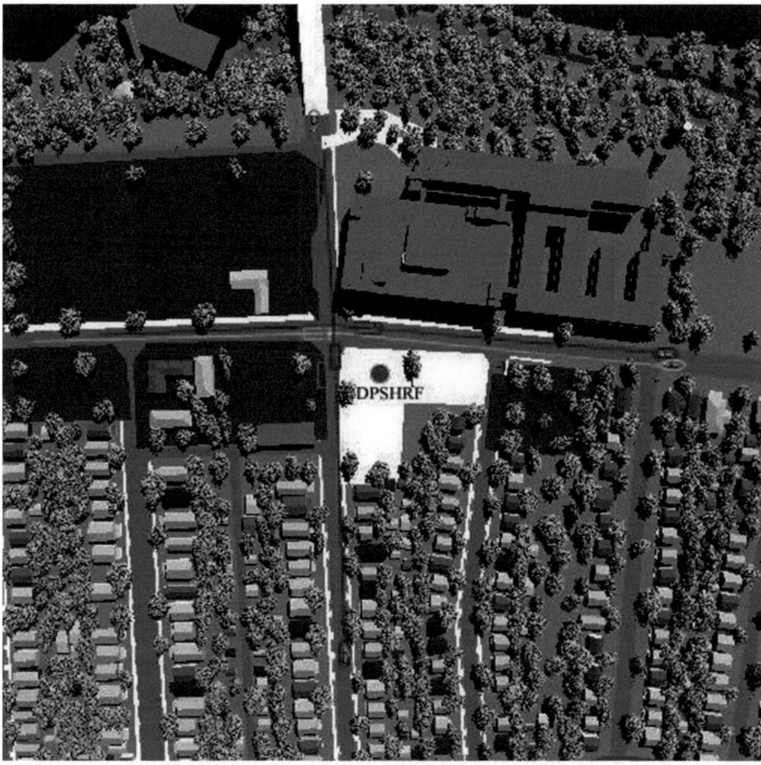

Fig. 2 A simulated urban scene image captured at latitude = 43.0° and longitude = 77.0°, 1,000 m above. The *ellipses* show the initial state of the four cars. The *rectangles* show the state where those cars move after 12 s. The "DPSHRF" sensor (in *solid dot*) is placed in front of the main large building. The simulated scene is captured at 8 am in a typical summer day

background, multiple moving targets, and the ego-motion of the sensor platform. The focal length of the hyperspectral imager is automatically adjusted according to the target detection results generated from the two line scanners. To simulate the hyperspectral imager, we use a frame array sensor with small spatial resolution at 70×70 for the hyperspectral data, and the ground truth range data provided by DIRSIG are transformed into range images. The spectral resolution is 0.01 μm ranged from 0.4 to 1.0 μm. Different portion of bandwidth can be selected and determined by analyzing the model spectral profile.

The simulation will enable a close investigation of intelligent sensor designs and hyperspectral data selection and exploitation for user designated targets. The DIRSIG simulation environment allows us to use an iterative approach to multimodal sensor designs. Starting with user and application requirements, various targets of interest in different, cluttered background can be simulated using the scene–target simulation tools in the DIRSIG. Then the adaptive multimodal sensor that has been designed can be modeled using the sensor modeling tools within the DIRSIG, and multimodal sensing data (images) can be generated. Target detection/identification, background modeling and multimodal fusion algorithms will be run on these simulated images to evaluate the overall performance of the automated target recognition, and to investigate the effectiveness of the initial multimodal sensor design. The evaluations of the recognition results against the given "ground-truth" data (by simulation) can provide further indicators for improving the initial sensor design, for example, spatial resolution, temporal sampling rates, spectral band selection, the role of range information and polarization, etc. Finally, a refined sensor design can again be modeled within the DIRSIG to start another iteration of sensor and system evaluation.

5 Data Exploitation and Adaptive Sensing

The basic procedure for active target detection and tracking is as the follows. A few selected spectral bands are used to initialize the detection of targets either based on motion detection or scene/target properties in prior scenarios. Then, for the potential interesting targets, the fovea turns to each of them to get a high-resolution, hyperspectral image with range information. This can be done in real-time so that tracking of one target and switching between multiple candidates is made possible. Finally, the signatures of the targets can be obtained by automatically analyzing the hyperspectral data in the fovea and by selecting the most relevant bands for such targets. This kind of function needs the active control of the sensor to fuse the peripheral and fovea vision in an efficient manner. In the following, we elaborate the principle by using some commonly used algorithms in target detection, tracking and identification, using our bio-inspired multimodal sensor.

5.1 Detection and Tracking in Peripheral Views

The first step is to find ROIs that possibly contain moving targets (Fig. 3). Simple background subtraction between a scanned image and a background image is not sufficient because the panoramic background (with trees, building, etc.) may change due to illumination changes over a large span of time. The advantage of using the two consecutive scanners is the ability to quickly detect a moving target in real time using "frame difference" without producing too much noise from the background. Further, a morphological noise removal technique [18] is applied to remove small sparse noises with the opening operation and fill small holes with the closing operation. However, the results from "frame difference" cannot provide accurate location and size information of the moving targets. Therefore, bounding boxes are defined from the "frame difference" results to mask off those background regions for background subtraction, which can provide more accurate location and size information of the moving targets. Figure 3c shows some bounding boxes that can be used as masks for performing the background subtraction of each individual panoramic scan.

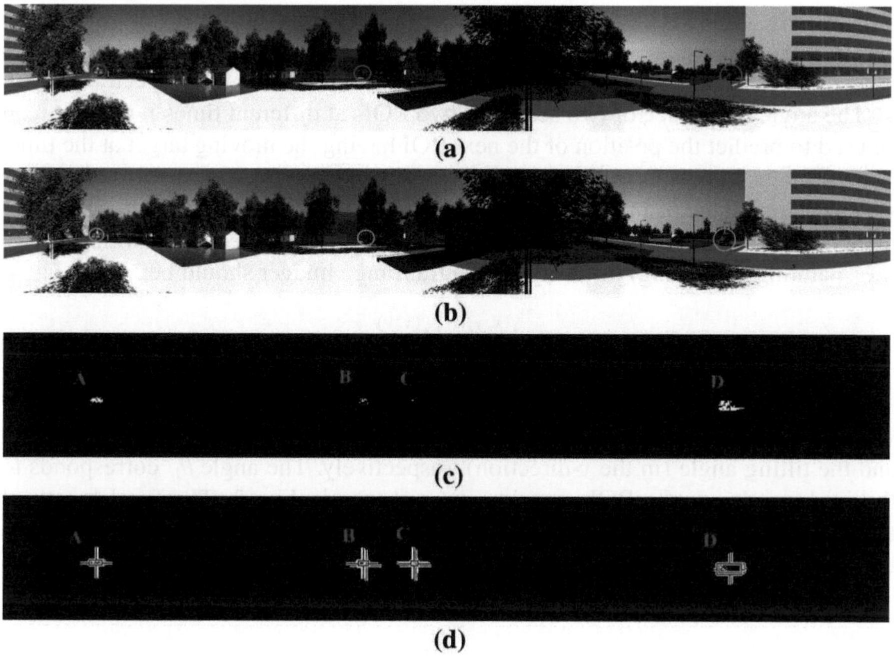

Fig. 3 All 360° panoramic images (512 × 3,600) shown here are integrated from vertical scan lines captured by the dual-panoramic scanners. **a** Panoramic image from the first scanner, with the moving targets indicated inside *circles*. **b** Panoramic image from the second scanner, again the same moving targets indicated inside *circles*. **c** Frame difference between **b** and **c**, group of ROIs are labeled. **d** Background subtraction from two scans inside boundaries defined by **c**. *rectangles* showed ROIs from two scans. (Close-up view of each labeled region can be seen clearly in Table 1.)

The threshold is set very low since we are interested in any changes in motion comparing to the relative static background. Of course, false alarms can also be generated by events such as the change of a large shadow, but this can be verified once we captured the hyperspectral image. The background image is updated for only those pixels belonging to the background after each 360° rotation, thus moving object extraction is maintained over time. At every rotation, each of the two line scanners will generate a sequence of 1D image lines that are combined to generate the panorama. Thus, registration problems can be avoided with the stabilized line scanners. Real-time target detection can be achieved since the scanning and detection are performed simultaneously and continuously.

The next step is to estimate the region of the next position that may contain a target once the two ROIs of the same target are found at two different times resulting from two different scans (Fig. 3d). The location and size differences of the two regions can determine the relative bearing angle of the hyperspectral/range fovea imager to zoom on the moving target. The position of extracted region from the dual-scans indicates which direction the target is moving to. Also, the size of two regions can indicate whether the target is moving closer to the sensor or farther. Therefore, we can calculate the next position where the target will be. Then, the ratio of the previous two regions can be used to estimate the new focal length of the hyperspectral imager.

The angle difference of two scans for two ROIs at different times t_i and t_{i+1} can be used to predict the position of the next ROI having the moving target at the time, t_{i+2}, when the hyperspectral/range imager can be in place. Therefore, given the time, we can estimate the panning and tile angles of the hyperspectral/range imager. Note that only the angles relative to the center of a region are needed. The turning angles (i.e., panning and tilting) of the hyperspectral/range imager should be:

$$\theta_{t_{i+2}}^{(x,y)} = \theta_{t_{i+1}}^{(x,y)} + \frac{t_{i+2}^{(x,y)} - t_{i+1}^{(x,y)}}{t_{i+1}^{(x,y)} - t_{i}^{(x,y)}} \left(\theta_{t_{i+1}}^{(x,y)} - \theta_{t_i}^{(x,y)}\right) \tag{1}$$

where the superscript x and y correspond to the panning angle (in the x-direction) and the tilting angle (in the y-direction), respectively. The angle θ_{t_i} corresponds to the angle position of a ROI at a time t_i as shown in Fig. 3. The focal length of the hyperspectral/range fovea is inversely proportional to the desired FOV of the hyperspectral/range imager, α, in order to have the target in the full view of the FOV. The FOV angle can be estimated as

$$\alpha = \frac{R_{t_{i+2}}}{P^l} \tag{2}$$

where $R_{t_{i+2}}$ is the predicted size of the target region at t_{i+2}, and P^l is the number of scanning lines per radius. The relationship between $R_{t_{i+2}}$ and the previous two regions of the same target at different times can be expressed as

$$\frac{R_{t_{i+1}}}{R_{t_i}}(t_{i+1} - t_i) = \frac{R_{t_{i+2}}}{R_{t_{i+1}}}(t_{i+2} - t_{i+1}) \tag{3}$$

Then a hyperspectral foveal shot of a ROI from the calculation can be taken. Thus, hyperspectral/range data is recorded in a more efficient way, only for ROIs. It is possible for some regions to be identified that do not have true moving targets inside. Then the hyperspectral classification in next step can verify this situation.

5.2 Target Classification Using 3D and HSI Fovea

Targets can be classified based on hyperspectral measurements, shape information, and the integration of both. There has been a lot of work in recognizing objects using 3D shape information (e.g., [3, 19]). Here we will only describe how to use a target's depth information and the information of its background to perform better hyperspectral classification.

Recognizing a target needs to compare the target's spectrum associated with each pixel to its training spectrum. In our experiments, a spectral library was pre-built with some existing models. Various vehicles with different colors and shapes can be imported and tested in the simulation scene. In the particular scenario in Fig. 2, four cars having the same shape but different paints are modeled. Two are red, one is brown and one is black. Initial spectral signatures of the four cars were captured from different angles in the same background. The capturing angles and surroundings are important and need to be considered carefully because those factors can significantly affect the effective radiance reaching the sensor, $L(l, \theta, \varphi, \lambda)$, where l is the slant range from sensor to target, θ, φ and λ are the zenith angle, the azimuth angle and the wavelength, respectively. The general expression for L is more complex and fully described in [17]. However, we can simplify L if we are only interested in the reflective (visible) bands, the general equation can be further expressed as:

$$L(l, \theta, \phi, \lambda) = f(L_s(l, \sigma, \lambda), L_{ds}(\theta, \phi), L_{bs}(\theta, \phi, \lambda), L_{us}(l, \theta, \lambda)) \qquad (4)$$

where σ is the angle from the normal to the target to the sun, L_s is the solar radiance, L_{ds} is the downwelled radiance from the sky due to the atmospheric scattering, L_{bs} is the spectral radiance due to the reflection from background objects, and L_{us} is the scattered atmospheric path radiance along the target-sensor line of site.

In the training stage, the background is known and fixed, thus L_{bs} can be cancelled out. The angles of the sun to the target and the of target to the sensor are known, thus we can keep this information and estimate a new spectral profile of the model target once we need to monitor a new target at a different time. L_{ds} and L_{us} can also affect the initial spectral profile if the weather condition changes significantly. In the current experiments, we only use one atmospheric dataset which can also be replaced and changed in the simulation in the future. After handling all reflective variants, various endmembers that represent the spectral extremes that best characterize a material type of a target were selected, and their spectral curves were stored in the spectral library database. We used the sequential maximum angle convex cone (SMACC) [8] to extract spectral endmembers and their abundance for every model target. In comparison to the conventional pixel purity index (PPI) [2] and N-FINDER [20], SMACC is a much faster and more automated method for finding spectral endmembers. Simply speaking, SMACC first finds extreme points or vectors that cannot be represented by a positive linear combination of other vectors in the data as a convex cone, and then a constrained oblique projection is applied to the existing cone to derive the next endmembers. The process is repeated until a tolerance value is reached, for example, max number of endmembers. Each endmember spectrum, defined as H, can be presented mathematically as a combination of the product of a convex 2D matrix contains endmember spectra as columns and a positive coefficient matrix:

$$H(c, i) = \sum_{k}^{N} R(c, k) A(k, j) \qquad (5)$$

where i is the pixel index, j and k are the endmember indices, and c is the spectra channel index. Some endmembers might have less spectra differences in term of redundancy. Those can be coalesced based on a threshold so that the most extreme spectra are identified and used to represent the entire coalesced group of endmembers.

In the testing stage, the same target spectra may be varied in different conditions such as various surface orientations and surroundings. However, the significant spectral signature of a target can be estimated and maybe further corrected with the help of range information produced from a ranger finder. Knowing the angles of the sun and the sensor, the depth map (i.e., range data) can indicate whether the information of a background object close to the target should be counted when processing the target spectra. The result spectra will have similar shape but the magnitudes will be still different due to the variations of illumination intensities and directions. A spectral angle mapper (SAM) [14] algorithm is used to match the target spectra to reference spectra. The SAM is insensitive to illumination and albedo effects. The algorithm determines the spectral similarity between two spectra by calculating the angle between the spectra and treating them as vectors in a space with dimensionality equal to the number of bands [14]. Smaller angles represent closer match. The depth information and the relative location of the sun and the sensor can determine whether a target's spectra should be adjusted by the surrounding spectra when performing classification. As a result, each pixel is classified either to a known object if the target spectrum is matched with the library spectrum of that object, or to an unknown object, for instance, the background. To distinct multiple objects from database, the results from different group of endmembers of different targets are compared.

5.3 Simulated Experimental Results

Table 1 shows the processed results for the following four cases. (A) Multiple targets with different spectral signature. (B) A target is under a shadow cast by trees. (C) There is no moving target (thus a false alarm). (D) Only one side of the target spectral signature can be acquired and the other side cannot be determined due to the insufficient reflectance of the sun light and the surroundings. At this stage, we only recognize if the detected target is the car or not the car. The target region may not fully match to the right shape of the car model. Only the sub-region with sample pixels spectra are selected for the matching. From scenarios A and B, both frontal and side shape of the car can be recognized. However, in scenario D, the side of the car cannot be detected due the shadow from the nearby building. We also captured multiple shots when those cars moved to various locations following the trajectories indicated in Fig. 2. We can recognize those cars with different colors, but most false targets are resulted from the large shadow. Various solutions can be possible, for example: (1) to place the sensor platform at another position; (2) to reconfigure sensor parameters such as adjust the height and the pointing direction; and (3) to implement a better classification algorithm. Therefore the experimental results can quickly drive feedback to adjust and improve the sensor design and the algorithm implementations. Various scenarios and cases can be constructed and tested in the simulation framework before a real sensor is even made.

Table 1 Processing results of the simulated urban scene

Index	ROIs	Fovea Parameters	Fovea Shot	Sample Spectral Profile	Spectral Curves Annotations	Depth Map	SAM no depth	SAM with depth	Results
A		Zenith: 89.0 Azimuth: 80.0 Focal Length: 245mm		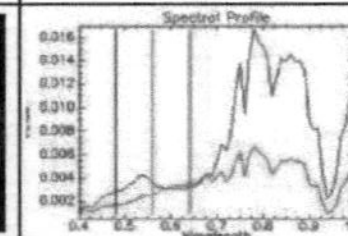	Top Curve: Car on Right Bottom Curve: Car on Left				Red Car Brown Car
B		Zenith: 88.5 Azimuth 191.0 Focal Length: 205mm		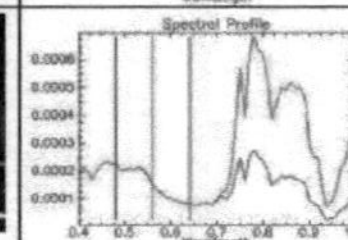	Top Curve: Car not in Shadow Bottom Curve: Car in Shadow		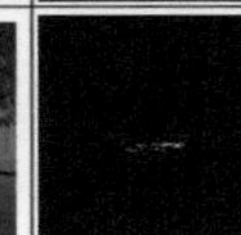		Red Car
C		Zenith: 88.5 Azimuth 220.0 Focal Length: 225mm	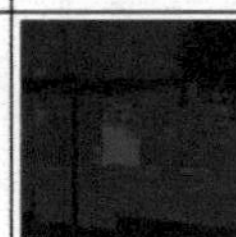	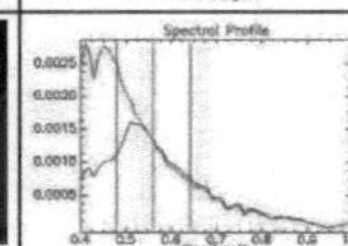	Top Curve: Material 1 Bottom Curve: Material 2	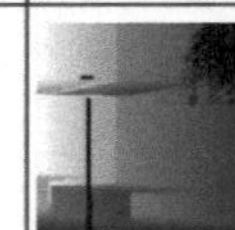			False Target
D	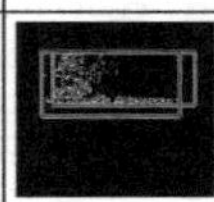	Zenith: 88.0 Azimuth 330.0 Focal Length: 125mm		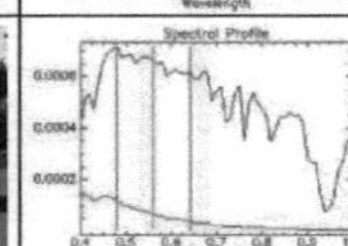	Top Curve: Front body Bottom Curve: Side body	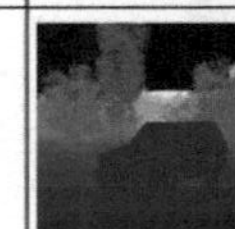		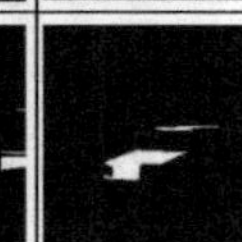	Black Car

Each index corresponds to each labeled region in Fiǵ. 3d. The column ROIs shows close-up view of result indicated in Fig. 3d. Hyperspectral fovea shots demonstrated here with only 3 RGB bands (which are also marked as vertical lines in the sample spectral profile column, in blue, green and red, respectively. Only the significant spectral signatures of targets are shown here. Final mapping results are shown in binary only to indicate the targets and the background. The classification is based on the match result with each model target spectral profile in database.

One of the useful advantages of the co-registered hyperspectral and range imaging is to using the range information to improve the effectiveness of the hyperspectral measurements. For example, in Table 1B, the shadowing of the vehicle (the red car) under the trees can be analyzed by the relation among the location of the sun, the locations of the trees from the panoramic background, and the surface orientations of the vehicle. Considering the depth information, the SAM can be obtained for surfaces of the vehicle under the influence of the tree shadows (therefore looks greenish). In Table 1D, the color only information is not sufficient to recognize the right target at where the background is also selected as the same one. With the depth information, the relations between the surfaces orientations of the vehicle (the black car) and the location of the sun can also tell which surfaces are illuminated. Therefore the well-illuminated surfaces (i.e., the top of the car body) can be selected based on the structural information obtained from the range data. The analysis so far is very preliminary but is very promising for future research.

6 Conclusion and Discussions

In this chapter, we first briefly described our system architecture and its characteristics accommodating with sensors design and algorithm. Then, we mainly described our bio-inspired multimodal sensor design that enables efficient hyperspectral data collection for tracking moving targets in real-time. This design and the related processing steps are tested through a system approach with sensor modeling, realistic scene simulation, and data exploitation. By simulation, various components can be reconfigured or replaced for specific situations or tasks. The image processing algorithms are designed only to demonstrate the basic idea of effectively capturing hyperspectral data in ROIs based on data exploitation. Needless to say, more sophisticated algorithms need to be developed for more challenging tasks. We only described one spectral classification method for recognizing the object. More precise and efficient hyperspectral classification routines may be applied. In addition, error characterizations of the hyperspectral sensing and range sensing have not been discussed. These are the standard procedures in image analysis and computer vision; our simulation approach will facilitate the simulation and evaluation of the system performance under various signal-to-noise ratios (SNRs). This remains our future work.

The real-time hyperspectral/range fovea imaging further extends the capability of human fovea vision. In the future, we will study two aspects of data processing: range-spectral integration and intelligent spectral band selection. Both issues will be greatly facilitated by our system approach and advanced scene and sensor simulation.

Range-spectral integration. There are many factors that need to be considered in correcting the acquired hyperspectral data to reveal the true material reflectance, including source illumination, scene geometry, atmospheric and sensor effects, spectral and space resolution, and etc. In the low-altitude airborne or ground imaging cases, the scene geometry is probably the most important factor. Therefore, the design of co-registered hyperspectral and range fovea will provide both spectral and geometry measurements of the 3D scene in a high resolution, so that a range-aided spectral correction can be performed. Using the DIRSIG tools, we have simulated both hyperspectral images and ranges images for several selected targets with known 3D models and spectral properties, and the next step to derive algorithms to perform spectral correction by the more effective 3D structure information of the targets given by the range images and the background information given by the panoramic scanners.

Optimal band selection. After the analysis of the hyperspectral data, the most useful wavelengths that can capture the target's signatures can be selected via tunable filtering; and the task of tracking and target recognition will only need to use the few selected bands or a few key features rather than all of the bands. This study will be carried out in several scenarios involving different targets in a challenging background or different backgrounds. We will compare the hyperspectral profiles (i.e., 3D images with two spatial dimensions and a spectral dimension) of various targets against different background materials, and then derive the optimal spectral signatures to distinguish a target from its background. We will also investigate how the range information can be used in improving the effectiveness of signature extraction and target recognition. The DIRSIG target and scene simulation tools could provide sufficient samples as training examples for us to optimal hyperspectral band selection.

On one hand, the design of a comprehensive system architecture such as DMPA stems from the requirements of managing a large-scale, multiple tasks, and distributed sensor systems. We made the first attempt to apply a system approach with a system architecture for multimodal sensor designs. However, the results are very preliminary and it is still a challenging issue to evaluate the usefulness of such an architecture for improving sensor designs. We hope our proposed idea will stir more research interests in looking into this problem. From our very early study, any multimodal sensor design that is implemented within the DPMA framework will have a number of characteristics. First, the sensor system model will have one or more sensing devices that receive stimulation from the environment. It is likely that these sensing devices operate independently and can have parameter settings controlled from system control programs. The data produced by various components should be reusable and be preserved to be compared with other results. Second, the system model will begin with a few components that monitor the behaviors of sensing and data processing, and then will grow over time as components are added and the structure is refined. All components models should be reusable. Third, simulation model can be used to describe observations which contain the state of components, the inputs and the response and actions at various times. At last, human interaction should be provided for designers not only to understand the interactions and evaluate performance of the system but also to instantiate different components, set their parameters, and, in general, prescribe all aspects of simulation.

Acknowledgments This work is supported by AFOSR under Award #FA9550-08-1-0199, and in part by AFRL/SN under Award No. FA8650-05-1-1853 and by NSF under Grant No. CNS-0551598.

References

1. Becklin, J.A.D.E.E., Casey, S.C., Savage, M.L.: Stratospheric observatory for infrared astronomy (sofia). In: Fowler, A. (ed.) Proceedings of SPIE, vol. 3356, p. 492. The International Society for Optical Engineering (1998)
2. Boardman, J.W., Druse, F.A., Green, R.O.: Mapping target signatures via partial unmixing of AVIRIS data. In: Fifth Annual JPL Airborne Earth Science Workshop, vol. 1, AVIRIS Workshop, pp. 23–26 (1995)
3. Diplaros, A., Geves, T., Patras, I.: Combining color and shape information for illumination-viewpoint invariant object recognition. Image Process **1**(1), 1–11 (2006)
4. DIRSIG http://dirsig.cis.rit.edu/. Last visited October (2008)
5. Fibreoptic Systems Inc. 60 Moreland Rd, Unit A, Simi Valley, CA 93065, USA http://www.fibopsys.com/
6. Fletcher-Holmes, D.W., Harvey, A.R.: Real-time imaging with a hyperspectral fovea. J. Opt. A Pure Appl. Opt. **7**, S298–S302 (2005)
7. Goldberg, A.C., Stann, B., Gupta, N.: Multispectral, hyperspectral, and three-dimensional imaging research at the U.S. Army research laboratory. In: Proceedings of the Sixth International Conference of Information Fusion, 2003, vol. 1, pp. 499–506 (2003)
8. Gruninger, J., Ratkowski, A.J., Hoke, M.L.: The sequential maximum angle convex cone (SMACC) endmember model. In: Proceedings of SPIE, Algorithms for Multispectral and Hyper-spectral and Ultraspectral Imagery, vol. 5425(1), Orlando, FL (2004)
9. Harvey, A.R., Fletcher-Holmes, D.W.: Imaging apparatus. GB Patent Application (0215)248.6 (2002)
10. Horn, S.: DARPA's Adaptive Focal Plane Array (AFPA) Program. Available: http://www.arpa.mil/mto/programs/afpa/index.html
11. Headwall Photonics Inc., 601 River Street, Fitchburg, MA 01420, USA http://www.headwallphotonics.com/
12. Horn, S.: DARPA's Adaptive Focal Plane Array (AFPA) Program. Available: http://www.arpa.mil/mto/programs/afpa/index.html. Last visited (2007)
13. Kincade, K.: MIRTHE center aims to take mid-IR sensors to new heights. (Optical Sensing). Laser Focus World (2006)
14. Kruse, F.A., et al.: The spectral image processing system (SIPS)—interactive visualization and analysis of imaging spectrometer data. Rem. Sens. Environ. **44**, 145–163 (1993)
15. Land, M.F., Nilsson, D.-E.: Animal Eyes. Oxford University Press, Oxford (2002)
16. Schott, J.R., Brown, S.D., Raqueño, R.V., Gross, H.N., Robinson, G.: An advanced synthetic image generation model and its application to multi/hyperspectral algorithm development. Can. J. Rem. Sens. **25**(2), 99–111 (1999)
17. Schott, J.R.: Remote Sensing: The Image Chain Approach, 2nd edn. Oxford University Press, Oxford (2007)
18. Soille, P.: Morphological Image Analysis: Principle and Applications. Springer, Berlin (1999)
19. Start, T.M., Fischler, M.A.: Context-based vision: recognizing objects using information from both 2D and 3D imagery. IEEE Trans. Pattern Anal. Mach. Intell. **13**(10), 1050–1065 (1991)
20. Winter, M.F.: N-FINDR: an algorithm for fast autonomous spectral end-member determination in hyperspectral data. In: Proceedings of SPIE, vol. 3753, pp. 266–275 (1999)

Part III
Hyperspectral Image Analysis

Affine Invariant Hyperspectral Image Descriptors Based upon Harmonic Analysis

Pattaraporn Khuwuthyakorn, Antonio Robles-Kelly and Jun Zhou

Abstract This chapter focuses on the problem of recovering a hyperspectral image descriptor based upon harmonic analysis. It departs from the use of integral transforms to model hyperspectral images in terms of probability distributions. This provides a link between harmonic analysis and affine geometric transformations between object surface planes in the scene. Moreover, the use of harmonic analysis permits the study of these descriptors in the context of Hilbert spaces. This, in turn, provides a connection to functional analysis to capture the spectral cross-correlation between bands in the image for the generation of a descriptor with a high energy compaction ratio. Thus, descriptors can be computed based upon orthogonal bases capable of capturing the space and wavelength correlation for the spectra in the hyperspectral imagery under study. We illustrate the utility of our descriptor for purpose of object recognition on a hyperspectral image dataset of real-world objects and compare our results to those yielded using an alternative.

NICTA is funded by the Australian Government as represented by the Department of Broadband, Communications and the Digital Economy and the Australian Research Council through the ICT Centre of Excellence program.
Pattaraporn Khuwuthyakorn would like to acknowledge the support of the Australian Government's Cooperative Research Centres Program.

P. Khuwuthyakorn (✉) · A. Robles-Kelly · J. Zhou
Research School of Information Sciences and Engineering (RSISE),
Australian National University, Canberra, ACT 0200, Australia
e-mail: u4420081@anu.edu.au

P. Khuwuthyakorn · A. Robles-Kelly · J. Zhou
National ICT Australia (NICTA), Canberra, ACT 2601, Australia

P. Khuwuthyakorn
Cooperative Research Centre for National Plant Biosecurity,
Bruce, ACT 2617, Australia

R. Hammoud et al. (eds.), *Machine Vision Beyond Visible Spectrum*,
Augmented Vision and Reality, 1, DOI: 10.1007/978-3-642-11568-4_8,

Keywords Hyperspectral image descriptor · Harmonic analysis · Heavy-tailed distributions

1 Introduction

With the advent and development of new sensor technologies, it is now possible to capture image data in tens or hundreds of wavelength-resolved bands covering a broad spectral range. Compared to traditional monochrome and trichromatic cameras, hyperspectral image sensors provide an information-rich representation of the spectral response for the material under study over a number of wavelengths. This has opened-up great opportunities and posed important challenges due to the high dimensional nature of the spectral data. As a result, many classical algorithms in pattern recognition and machine learning have been naturally borrowed and adapted so as to perform feature extraction and classification [20]. Techniques such as Principle Component Analysis (PCA) [17], Linear Discriminant Analysis (LDA) [12], Projection Pursuit [16] and their kernel versions [10] treat raw pixel spectra as input vectors in a higher-dimensional space, where the dimensionality is given by the number of bands. The idea is to recover statistically optimal solutions to the classification problems by reducing the data dimensionality via a projection of the feature space.

The methods above are often used for purposes of recognition based on individual signatures, which in hyperspectral images, represent single pixels. Nonetheless each signature is generally related to material chemistry, these methods do not take into account the local structure of the images under study. They rather hinge in the notion that different materials have different characteristic responses as a function of wavelengths which can be used to provide descriptions of the target objects. Thus, raw pixels are often treated as input vectors in high dimensional spaces.

In contrast with the pixel-based methods in hyperspectral imaging, the approaches available for content-based image retrieval often take into account the local structure of the scene. These methods often represent images as a bag of features so as to match query images to those in the database by computing distances between distributions of local descriptors. As a result, trichromatic object and image retrieval and classification techniques [6, 25, 37] are often based upon the summarisation of the image dataset using a codebook of visual words [22, 24, 29].

It is surprising that despite the widespread use of higher-level features for recognition and retrieval of monochromatic and trichromatic imagery, local hyperspectral image descriptors are somewhat under-researched. The use of local image descriptors opens-up great opportunities in recognition and classification tasks. Moreover, the multidimensional nature of local image features and descriptors may be combined to improve performance. For instance, Varma and Ray [41] have used a kernel learning approach to learn the trade-off between discriminative power and invariance of image descriptors in classification tasks. Other methods, such as the one in [5], rely upon clustering algorithms to provide improved organisation of the codebook. Other alternatives tend to view the visual words as multidimensional data and, making use of unsupervised learning, exploit similarity information in a graph-theoretic setting. Examples of these are the method presented by Sengupta and Boyer [34] and that developed by Shokounfandeh et al. [36], which employ information-theoretical criteria to hierarchically structure the dataset under study and pattern recognition methods to match the candidates.

Amongst local image descriptors, texture has not only found applications as a shape queue [13, 40], but also attracted broad attention for recognition and classification tasks [31]. Moreover, from the shape modelling perspective, static texture planes can be recovered making use of the structural analysis of predetermined texture primitives [1, 15, 18]. This treatment provides an intuitive geometrical meaning to the task of recovering the parameters governing the pose of the object by making use of methods akin to 3D view geometry. For dynamic textures, Sheikh et al. [35] have developed an algorithm for recovering the affine geometry making use of motion-magnitude constraints. Péteri and Chetverikov [27] have characterised dynamic textures using features extracted using normal flows. This builds on the comparative study in [11]. Ghanem and Ahuja [14] have used the Fourier phase to capture the global motion within the texture. Rahman and Murshed [30] have estimated optical flow making use of motion patterns for temporal textures. Otsuka et al. [26] have used surface motion trajectories derived from multiple frames in a dynamic texture to recover spatiotemporal texture features.

As mentioned earlier, we focus on the problem of recovering a hyperspectral image descriptor by using harmonic functions to model hyperspectral imagery in terms of probability distributions. This is reminiscent of time-dependent textures, whose probability density functions exhibit first and second order moments which are space and time-shift invariant [9]. For instance, in [28], the characterisation of the dynamic texture under study is obtained using the empirical observations of statistical regularities in the image sequence. In [3], statistical learning is used for purposes of synthesising a dynamic texture based upon an input image sequence. Zhao and Pietikäinen [43] have performed recognition tasks using local binary patterns that fit space–time statistics.

The methods above view time-dependent textures as arising from second-order stationary stochastic processes such as moving tree-leaves, sea waves and rising smoke plumes. We, from another point of view, relate hyperspectral image regions to harmonic kernels to capture a discriminative and descriptive representation of the scene. This provides a principled link between statistical approaches and signal processing methods for texture recognition to shape modeling approaches based upon measures of spectral distortion [23]. The method also provides a link to affine geometric transformations between texture planes and their analysis in the Fourier domain [4].

The chapter is organised as follows. We commence by exploring the link between harmonic analysis and heavy tailed distributions. We then explore the relationship between distortions over locally planar patches on the object surface and the domain induced by an integral transform over a harmonic kernel. We do this so as to achieve invariance to affine transformations on the image plane. With these technical foundations at hand, we proceed to present our hyperspectral image descriptor by incorporating the cross-correlation between bands. This results in a descriptor based upon orthogonal bases with high information compaction properties which can capture the space and wavelength correlation for the spectra in hyperspectral images. Moreover, as we show later on, the choice of bases or kernel is quite general since it applies to harmonic kernels which span a Hilbert space. We conclude the chapter by demonstrating the utility of our descriptor for purposes of object recognition based upon real-world hyperspectral imagery.

2 Heavy-Tailed Distributions

As mentioned earlier, we view hyperspectral images as arising from a probability distribution whose observable or occurrences may have long or heavy tails. This implies that the spectra in the image results in values that can be rather high in terms of their deviation from the image-spectra mean and variance. As a result, our formulation can capture high wavelength-dependent variation in the image. This is important, since it allows us to capture information in our descriptor that would otherwise may be cast as the product of outliers. Thus, we formulate our descriptor so as to model "rare" stationary wavelength-dependent events on the image plane.

Moreover, we view the pixel values of the hyperspectral image as arising from stochastic processes whose moment generating functions are invariant with respect to shifts in the image-coordinates. This means that the mean, covariance, kurtosis, etc. for the corresponding joint probability distribution are required to be invariant with respect to changes of location on the image. Due to our use of heavy tailed distributions, these densities may have high dispersion and, thus, their probability density functions are, in general, governed by further-order moments. These introduces a number of statistical "skewness" variables that allow modeling high variability spectral behaviour.

This is reminiscent of simulation approaches where importance sampling cannot be effected via an exponential changes in measurement due to the fact that the moments are not exponential in nature. This applies to distributions such as the log-normal, Weibull with increasing skewness and regularly varying distributions such as Pareto, stable and log-gamma distributions [2]. More formally, we formulate the density of the pixel-values for the wavelength λ at the pixel u in the image-band I_λ of the image as random variables $\mathscr{Y}_u$ whose inherent basis $\mathscr{X}_u = \{x_u(1), x_u(2), \ldots, x_u(|\mathscr{X}_u|)\}$ is such that

$$P(\mathscr{Y}_u) = \sum_{k=1}^{|\mathscr{X}_u|} P(x_u(k)) \tag{1}$$

where, $x_u(k)$ are identically distributed variables and, as usual for probability distributions of real-valued variables, we have written $P(\mathscr{Y}_u) = \Pr[y \leq \mathscr{Y}_u]$ for all $y \subset \mathfrak{R}$.

In other words, we view the pixel values for each band in the image under study as arising from a family of heavy-tailed distributions whose variance is not necessarily finite. It is worth noting that, for finite variance, the formalism above implies that $P(\mathcal{Y}_u)$ is normally distributed and, as a result, our approach is not exclusive to finite variance distributions, but rather this treatment generalises the stochastic process to a number of independent influences, each of which is captured by the corresponding variable $x_u(k)$.

In practice, the Probability Density Function (PDF) $f(\mathcal{Y}_u)$ is not available in close form. As a result, we can re-parameterise the PDF by recasting it as a function of the variable ς making use of the characteristic function

$$\psi(\varsigma) = \int_{-\infty}^{\infty} \exp(\mathbf{i}\varsigma\mathcal{Y}_u) f(\mathcal{Y}_u)\, d\mathcal{Y}_u \tag{2}$$

$$= \exp(\mathbf{i}u\varsigma - \gamma|\varsigma|^{\alpha}(1 + \mathbf{i}\beta\, \mathrm{sign}(\varsigma)\varphi(\varsigma, \alpha))) \tag{3}$$

where $i = \sqrt{-1}$, u is, as before, the pixel-index on the image plane and $\gamma \in \Re^{+}$ are function parameters, $\beta \in [-1, 1]$ and $\alpha \in (0, 2]$ are the skewness and characteristic exponent, respectively, and $\varphi(\cdot)$ is defined as follows:

$$\varphi(\varsigma, \alpha) = \begin{cases} \tan(\alpha\frac{\pi}{2}) & \text{if } \alpha \neq 1 \\ -\frac{\pi}{2}\log|\varsigma| & \text{if } \alpha = 1 \end{cases} \tag{4}$$

For the characteristic function above, some values of α correspond to special cases of the distribution. For instance, $\alpha = 2$ implies a normal distribution, $\beta = 0$ and $\alpha = 1$ corresponds to a Cauchy distribution and, for the Levy distribution we have $\alpha = \frac{1}{2}$ and $\beta = 1$. Thus, nonetheless the formalism above can capture a number of cases in exponential families, it is still quite general in nature so as to allow the modeling of a large number of distributions that may apply to hyperspectral data and whose characteristic exponents α are not of those distributions whose tails are exponentially bounded.

So far, we have limited ourselves to the image plane for a fixed wavelength λ. That is, we have, so far, concentrated on the distribution of spectral values across every wavelength-resolved band in the image. Note that, without loss of generality, we can extend Eq. 3 to the wavelength domain, i.e., the spectra of the image across a segment of bands.

This is a straightforward task by noting that the equation above can be viewed as the cross-correlation between the function $f(\mathcal{Y}_u)$ and the exponential given by $\exp(\mathbf{i}\varsigma\mathcal{Y}_u)$. Hence, we can write the characteristic function for the image parameterised with respect to the wavelength λ as follows:

$$\vartheta(\lambda) = \int_{-\infty}^{\infty}\int_{-\infty}^{\infty} \exp(\mathrm{i}\lambda\varsigma)\exp(\mathrm{i}\varsigma\mathcal{Y}_u) f(\mathcal{Y}_u)\, d\mathcal{Y}_u\, d\varsigma \tag{5}$$

$$= \int_{-\infty}^{\infty} \exp(\mathrm{i}\lambda\varsigma)\psi(\varsigma)\, d\varsigma \tag{6}$$

where the second line in the equation above corresponds to the substitution of Eq. 3 into Eq. 5.

Equation 6 captures the spectral cross-correlation for the characteristic functions for each band. In this manner, we view the characteristic function for the hyperspectral image as a heavy-tailed distribution of another set of heavy-tailed PDFs, which correspond to each of the band in the image. This can also be interpreted as a composition of two heavy-tailed distributions, where Eq. 3 corresponds to the image-band domain ς of the image and Eq. 6 is determined by the wavelength-dependent domain λ.

This composition operation suggests a two-step process for the computation of the image descriptor. Firstly, at the band-level, the information can be represented in a compact fashion making use of harmonic analysis and rendered invariant to geometric distortions on the object surface plane. Secondly, the wavelength-dependent correlation between bands can be computed making use of the operation in Eq. 6.

3 Harmonic Analysis

In this section, we explore the use of harmonic analysis and the fundamentals of integral transforms [38] to provide a means to the computation of our image descriptor. We commence by noting that Eqs. 2 and 5 are characteristic functions obtained via the integral of the product of the function $g(\eta)$, i.e., $f(\mathscr{Y}_u)$ and $\psi(\varsigma)$, multiplied by a kernel, given by $\exp(\mathrm{i}\lambda\varsigma)$ and $\exp(\mathrm{i}\varsigma\mathscr{Y}_u)$, respectively.

To appreciate this more clearly, consider the function given by

$$F(\omega) = \int_{-\infty}^{\infty} g(\eta) K(\omega, \eta)\, d\eta \tag{7}$$

where $K(\omega, \eta)$ is a harmonic kernel of the form

$$K(\omega, \eta) = \sum_{k=1}^{\infty} \mathfrak{a}_k \phi_k(\omega) \phi_k(\eta) \tag{8}$$

where $\mathfrak{a}_k$ is the kth real scalar corresponding to the harmonic expansion and $\phi_k(\cdot)$ are orthonormal functions such that $\langle \phi_k(\omega), \phi_n(\eta) \rangle = 0 \ \forall \ n \neq k$. Moreover, we consider cases in which the functions $\phi_k(\cdot)$ constitute a basis for a Hilbert space [42] and, therefore, the right-hand side of Eq. 8 is convergent to $K(\omega, \eta)$ as k tends to infinity.

To see the relation between Eq. 7 and the equations in previous sections, we can examine $\psi(\varsigma)$ in more detail and write

$$\log[\psi(\varsigma)] = \mathrm{i}u\varsigma - \gamma|\varsigma|^{\alpha}(1 + \mathrm{i}\beta \operatorname{sign}(\varsigma)\psi(\varsigma, \alpha)) \tag{9}$$

$$= \mathrm{i}u\varsigma - |\varsigma|^{\alpha}\gamma^{*\alpha} \exp\left(-\mathrm{i}\beta^{*}\frac{\pi}{2}\vartheta \operatorname{sign}(\varsigma)\right) \tag{10}$$

where $\vartheta = 1 - |1 - \alpha|$ and parameters γ^* and β^* are given by

$$\gamma^* = \left(\frac{\gamma \sqrt{\Omega}}{\cos(\alpha \frac{\pi}{2})} \right)^{\frac{1}{\alpha}} \tag{11}$$

$$\beta^* = \frac{2}{\pi \vartheta} \arccos \left(\frac{\cos(\alpha \frac{\pi}{2})}{\sqrt{\Omega}} \right) \tag{12}$$

and $\Omega = \cos^2(\alpha \frac{\pi}{2}) + \beta^2 \sin^2(\alpha \frac{\pi}{2})$.

To obtain the kernel for Eq. 7, we can use Fourier inversion on the characteristic function and, making use of the shorthands defined above, the PDF may be computed via this following equation.

$$f(\mathcal{Y}_u; u, \beta^*, \gamma^*, \alpha) = \frac{1}{\pi \gamma^*} \int_0^{\infty} \cos \left(\frac{(u - \mathcal{Y}_u)s}{\gamma^*} + s^{\alpha} \sin(\phi) \right) \exp(-s^{\alpha} \sin(\phi))\, ds \tag{13}$$

where $\phi = \frac{\beta^* \pi \eta}{2}$.

This treatment not only opens-up the possibility of functional analysis on the characteristic function using the techniques in the Fourier domain, but also allows the use of other harmonic kernels for compactness and ease of computation. This is due to the fact that, we can view the kernel $K(\omega, \eta)$ as the exponential $\exp(-s^{\alpha} \sin(\phi))$, whereas the function $g(\eta)$ is given by the cosine term. Thus, we can use other harmonic kernels so as to induce a change of basis without any loss of generality. Actually, the expression above can be greatly simplified making use of the shorthands $A = \frac{(u - \mathcal{Y}_u)}{\gamma^*}$, $\eta = s^{\alpha}$ and $\omega\eta = As + s^{\alpha} \sin(\phi)$, which yields

$$s^{\alpha} \sin(\phi) = \omega\eta - A\eta^{\frac{1}{\alpha}} \tag{14}$$

Substituting Eq. 13 with Eq. 14, the PDF can be expressed as

$$f(\mathcal{Y}_u; u, \beta^*, \gamma^*, \alpha) = \sqrt{\frac{2}{\pi}} \int_0^{\infty} \frac{\exp(-\omega\eta + A\eta^{\frac{1}{\alpha}})}{\sqrt{2\pi}\gamma^* \alpha \eta^{\left(\frac{\alpha-1}{\alpha}\right)}} \cos(\omega\eta)\, d\eta \tag{15}$$

where the kernel then becomes

$$K(\omega, \eta) = \cos(\omega\eta) \tag{16}$$

This can be related, in a straightforward manner, to the Fourier cosine transform (FCT) of the form

$$F(\omega) = \sqrt{\frac{2}{\pi}} \int_0^{\infty} \frac{\exp\left(-\omega\eta + \frac{(u - \mathcal{Y}_u)}{\gamma^*} \eta^{\frac{1}{\alpha}} \right)}{\sqrt{2\pi}\gamma^* \alpha \eta^{\left(\frac{\alpha-1}{\alpha}\right)}} \cos(\omega\eta)\, d\eta \tag{17}$$

which is analogous to the expression in Eq. 13. Nonetheless, the transform above does not have imaginary coefficients. This can be viewed as a representation in the power rather than in the phase spectrum. Moreover, it has the advantage of compacting the spectral information in the lower-order Fourier terms, i.e., those for which ω is close to the origin. This follows the strong "information compaction" property of FCTs introduced in [32] and assures a good trade-off between discriminability and complexity.

It is worth stressing that, due to the harmonic analysis treatment given to the problem in this section, other kernels may be used for purposes of computing other integral transforms [38] spanning Hilbert Spaces. These include wavelets and the Mellin ($K(\omega, \eta) = \eta^{\omega-1}$) and Hankel transforms. In fact, other Kernels may be obtained by performing an appropriate substitution on the term $\cos(\omega\eta)$. Note that, for purposes of our descriptor recovery, we will focus on the use of the cosine transform above. This is due to the information compaction property mentioned earlier and the fact that computational methods for the efficient recovery of the FCT are readily available.

4 Invariance to Affine Distortions

Having introduced the notion of the harmonic analysis and shown how the probability density function can be recovered using a Fourier transform, we now focus on relation between distortions on the object surface plane and the Fourier domain. To this end, we follow [4] and relate the harmonic kernel above to affine transformations on the object locally planar shape. As mentioned earlier, the function $f(\mathscr{Y}_u)$ corresponds to the band-dependent component of the image and, as a result, its prone to affine distortion. This hinges in the notion that a distortion on the object surface will affect the geometric factor for the scene, but not its photometric properties. In other words, the material index of refraction, roughness, etc. remains unchanged, whereas the geometry of the reflective process does vary with respect to affine distortions on the image plane. The corresponding 2D integral transform of the function $f(\mathscr{Y}_u)$ which, as introduced in the previous sections, corresponds to the pixel values for the image-band I_λ in the image under study is given by

$$F(\xi) = \int_{\Gamma} f(\mathscr{Y}_u) K\left(\xi^T, u\right) du \tag{18}$$

where $u = [x, y]^T$ is the vector of two-dimensional coordinates for the compact domain $\Gamma \in \Re^2$ and, in the case of the FCT, $K(\xi^T, u) = \cos(2\pi(\xi^T u))$.

In practice, the coordinate-vectors u will be given by discrete quantities on the image lattice. For purposes of analysis, we consider the continuous case and note that the affine coordinate transformation can be expressed in matrix notation as follows.

$$u' = \begin{bmatrix} x' \\ y' \end{bmatrix} = \begin{bmatrix} a & b \\ d & e \end{bmatrix} \begin{bmatrix} x \\ y \end{bmatrix} + \begin{bmatrix} c \\ f \end{bmatrix} \tag{19}$$

This observation is important because we can relate the kernel for the FCT in Eq. 18 to the transformed coordinate $u' = [x', y']^T$. Also, note that, for patches centered at keypoints in the image, the locally planar object surface patch can be considered devoid of translation. Thus, we can set $c = f = 0$ and write

$$\xi^T u = \xi^T \begin{bmatrix} x \\ y \end{bmatrix} \tag{20}$$

$$= \begin{bmatrix} \xi_x \ \xi_y \end{bmatrix} \begin{bmatrix} a & b \\ d & e \end{bmatrix}^{-1} \begin{bmatrix} x' \\ y' \end{bmatrix} \tag{21}$$

$$= \frac{1}{ae - bd} \begin{bmatrix} (e\xi_x - d\xi_y) \ (-b\xi_x + a\xi_y) \end{bmatrix} \begin{bmatrix} x' \\ y' \end{bmatrix} \tag{22}$$

where $\xi = [\xi_x, \xi_y]^T$ is the vector of spectral indexes for the 2D integral transform.

Hence, after some algebra, and using the shorthand $\Delta = (ae - bd)$, we can show that for the coordinates u', the integral transform is given by

$$F(\xi) = \frac{1}{|\Delta|} \int_{-\infty}^{\infty} \int_{-\infty}^{\infty} f(\mathcal{Y}_{u'}) K \left(\frac{1}{\Delta} [(e\xi_x - d\xi_y), (b\xi_x - a\xi_y)], [x', y']^T \right) dx' dy' \tag{23}$$

This implies that

$$F(\xi) = \frac{1}{|\Delta|} F(\xi') \tag{24}$$

where ξ' is the "distorted" analogue of ξ. The distortion matrix $\mathbb{T}$ is such that

$$\xi = \begin{bmatrix} \xi_x \\ \xi_y \end{bmatrix} = \begin{bmatrix} a & d \\ b & e \end{bmatrix} \begin{bmatrix} \xi'_x \\ \xi'_y \end{bmatrix} = \mathbb{T}\xi' \tag{25}$$

As a result, from Eq. 23, we can conclude that the effect of the affine coordinate transformation matrix $\mathbb{T}$ is to produce a distortion equivalent to $(\mathbb{T}^T)^{-1}$ in the ξ domain for the corresponding integral transform. This observation is an important one since it permits achieving invariance to affine transformations on the locally planar object surface patch. This can be done in practice via a ξ-domain distortion correction operation of the form

$$F(\xi) = (\mathbb{T}^T)^{-1} F(\xi') \tag{26}$$

5 Descriptor Construction

With the formalism presented in the previous sections, we now proceed to elaborate further on the descriptor computation. Succinctly, this is a two-step process. Firstly, we compute the affine-invariant 2D integral transform for every band in the hyperspectral image under study. This is equivalent to computing the band-dependent component of the characteristic function $\psi(\varsigma)$. Secondly, we capture the wavelength dependent behaviour of the hyperspectral image by computing the cross-correlation with respect to the spectral domain for the set of distortion-invariant integral transforms. By making use of the FCT kernel, in practice, the descriptor becomes an FCT with respect to the band index for the cosine transforms corresponding to wavelength-resolved image in the sequence.

Following the rationale above, we commence by computing the distortion invariant integral transform for each band in the image. To do this, we use Eq. 26 to estimate the distortion matrix with respect to a predefined reference. Here, we employ the peaks of the power spectrum and express the relation of the integral transforms for two locally planar image patches, i.e., the one corresponding to the reference and that for the object under study. We have done this following the notion that a blob-like shape composed of a single transcendental function on the image plane would produce two peaks in the Fourier domain. That is, we have set, as our reference, a moment generating function arising from a cosine on a plane perpendicular to the camera.

Let the peaks of the power spectrum for two locally planar object patches, **A** and **B**, be given by $\mathbf{U}_A$ and $\mathbf{U}_B$. Those for the reference **R** are $\mathbf{U}_R$. The affine distortion matrices are $\mathbb{T}_A$ and $\mathbb{T}_B$ respectively. As a result, the matrices $\mathbf{U}_A$, $\mathbf{U}_B$ and $\mathbf{U}_R$ are such that each of their columns correspond to the x–y coordinates for one of the two peaks in the power spectrum. These relations are given by

$$\mathbf{U}_A = (\mathbb{T}_A^T)^{-1}\mathbf{U}_R \tag{27}$$

$$\mathbf{U}_B = (\mathbb{T}_B^T)^{-1}\mathbf{U}_R \tag{28}$$

where $\mathbb{T}_A : \mathbf{R} \Rightarrow \mathbf{A}$ and $\mathbb{T}_B : \mathbf{R} \Rightarrow \mathbf{B}$ are the affine coordinate transformation matrices of the planar surface patches under consideration in spatial domain.

Note that, this is reminiscent of the shape-from-texture approaches hinging in the use of the Fourier transform for the recovery of the local distortion matrix [33]. Nonetheless, in [33], the normal is recovered explicitly making use of the Fourier transform, whereas here we employ the integral transform and aim at relating the FCTs for the two locally planar patches with that of the reference. We can do this making use of the composition operation given by

$$\mathbf{U}_B = (\mathbb{T}_A\mathbb{T}_B^{-1})^T\mathbf{U}_A \tag{29}$$

$$\mathbf{U}_B = \Phi\mathbf{U}_A \tag{30}$$

where $\Phi = (\mathbb{T}_A\mathbb{T}_B^{-1})^T$ is the matrix representing the distortion of the power spectrum of $\mathbf{U}_A$ with respect to $\mathbf{U}_B$. Here, we use the shorthands ${\mathbb{T}_A}^T = \mathbf{U}_R{\mathbf{U}_A}^{-1}$ and $({\mathbb{T}_B}^T)^{-1} = \mathbf{U}_B{\mathbf{U}_R}^{-1}$ to write

$$\Phi = (\mathbf{U}_R{\mathbf{U}_A}^{-1})(\mathbf{U}_B{\mathbf{U}_R}^{-1}) \tag{31}$$

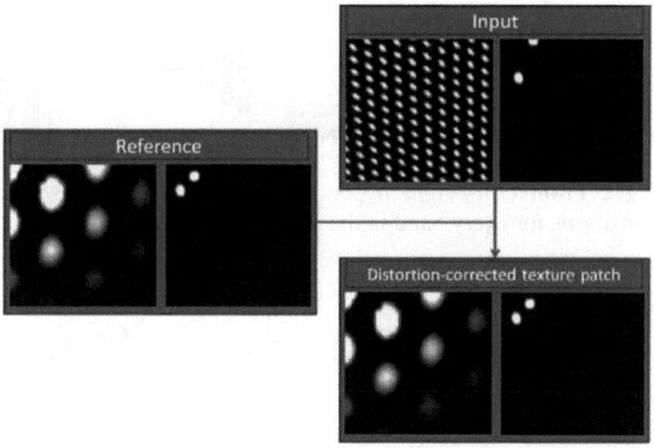

Fig. 1 Example of reference, input and distortion corrected single-band textures. In the panels, the *left-hand* image shows the single-band reference texture whereas the *right-hand* panel shows the power spectrum of the distorted and affine corrected FCT for the texture under study

As a result, we fix a reference for every locally planar patch so as to compute the matrix Φ directly through the expression above. This contrasts with other methods in the fact that, for our descriptor computation, we do not recover the principal components of the local distortion matrix, but rather construct a band-level matrix of the form

$$\mathbf{V} = [F(I_1)^*|F(I_2)^*|\ldots|F(I_{|\mathbb{I}|})^*] \tag{32}$$

which is the concatenation of the affine invariant integral transforms $F(\cdot)^*$ for the band-resolved locally planar object surface patches in the image. Moreover, we render the band-level integral transform invariant to affine transformations making use of the reference peak matrix $\mathbf{U}_R$ such that the transform for the band indexed t is given by

$$F(I_R) = F(I_t)^* {\Phi_t}^{-1} \tag{33}$$

where ${\Phi_t}^{-1}$ is the matrix which maps the transform for the band corresponding to the wavelength λ to the transform $F(I_R)$ for the reference plane. Here, as mentioned earlier, we have used as reference the power spectrum given by two peaks rotated 45° about the upper left corner of the 2D FCT. The reference FCT is shown in Fig. 1.

Note that, since we have derived our descriptor based upon the properties of integral transforms and Hilbert spaces, each element of the matrix $\mathbf{V}$ can be considered as arising from the inner product of a set of orthonormal vectors. Moreover, from a harmonic analysis perspective, the elements of $\mathbf{V}$ are represented in terms of discrete wave functions, over an infinite number of elements [19]. This is analogue to the treatment given to time series in signal processing, where the variance of the signal is described based upon spectral density. Usually, the variance estimations are performed by using Fourier transform methods [39]. Thus, we can make use of the discrete analogue of Eq. 6 so as to recover the kth coefficient for the image descriptor $\mathfrak{G}$, which becomes

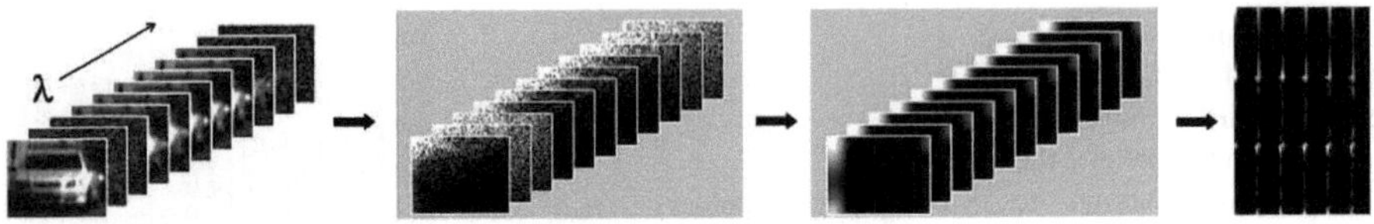

Fig. 2 From *left-to-right*: hyperspectral texture, the band-wise FCT, the distortion invariant cosine transforms for every band in the image and the raster scanned 3D matrix **V**

$$\mathfrak{G}_k = F(\mathbf{V}) = \sum_{n=0}^{|\mathbb{I}|-1} F(I_n)^* K\left(\frac{\pi}{|\mathbb{I}|}\left(n+\frac{1}{2}\right), \left(k+\frac{1}{2}\right)\right) \tag{34}$$

where $|\mathfrak{G}| = |\mathbb{I}|$ and, for the FCT, the harmonic kernel above becomes

$$K\left(\frac{\pi}{|\mathbb{I}|}\left(n+\frac{1}{2}\right), \left(k+\frac{1}{2}\right)\right) = \cos\left(\frac{\pi}{|\mathbb{I}|}\left(n+\frac{1}{2}\right)\left(k+\frac{1}{2}\right)\right) \tag{35}$$

6 Implementation Issues

Now, we turn our attention to the computation of the descriptor and provide further discussion on the previous developments. To this end, we illustrate, in Fig. 2, the step-sequence of the descriptor computation procedure. We depart from a series of bands in the image and compute the band-by-band FCT. With the band FCTs at hand, we apply the distortion correction approach presented in the previous sections so as to obtain a "power-aligned" series of cosine transforms that can be concatenated into **V**. The descriptor is then given by the cosine transform of **V** over the wavelength-index. Note that the descriptor will be three-dimensional in nature, with size $N_x \times N_y \times N_\lambda$, where N_x and N_y are the sizes of the locally planar object patches in the image lattice and N_λ is equivalent to the wavelength range for the hyperspectral image bands. In the figure, for purposes of visualisation, we have raster-scanned the descriptor so as to display a 2D matrix whose rows correspond to the wavelength-indexes of the hyperspectral image under study.

We now illustrate the distortion correction operation at the band level in Fig. 1. In the panels, we show the reference, corrected and input image regions in their spatial and frequency domains. Note that, at input, the textured planes show an affine distortion which affects the distribution of the peaks in its power spectrum.

Moreover, in Fig. 3, we show a sample textured plane which has been affinely distorted. In the figure, we have divided the distorted input texture into patches that are assumed to be locally planar. We then apply the FCT to each of these patches, represented in the form of a lattice on the input image in the left-hand panel. The corresponding power spectrums are shown in the second column of the figure. Note

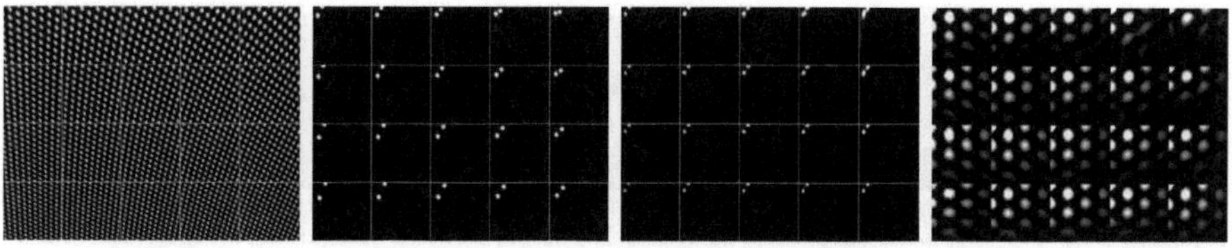

Fig. 3 From *left-to-right*: affine distortion of a sample single-band image; FCT of the image patches in the *left-hand panel*, distortion-corrected power spectrums for the FCTs in the *second panel* and inverse FCTs for the power spectrum in the *third panel*

that, as expected, the affine distortions produce a displacement on the power spectrum peaks. In the third panel, we show the power spectrums after the matrix Φ has been recovered and multiplied so as to obtain the corrected FCTs given by $F(\cdot)^*$. The distortion corrected textures in the spatial domain are shown in the right-most panel in the figure. These have been obtained by applying the inverse cosine transform to the power spectrums in the third column. Note that, from both, the corrected power spectrums and the inverse cosine transforms, we can conclude that the correction operation can cope with large degrees of shear in the input texture-plane patches.

7 Experiments

Having presented our image descriptor in the previous sections, we now illustrate its utility for purposes of hyperspectral image categorisation. To this end, we employ a dataset of hyperspectral imagery acquired in-house using an imaging system comprised by an Acousto-Optic Tunable Filter (AOTF) fitted to a firewire camera. The system has been designed to operate in the visible and near infrared (NIR) spectral range.

In our dataset, we have images corresponding to five categories of toys and a set of coins. Each toy sample was acquired over ten views by rotating the object in increments of 10° about its vertical axis whereas coin imagery was captured only in two different views, heads and tails. Figure 4 shows sample images over ten views of an object in the data set. In our database, there are a total of 62 toys and 32 coins, which, over multiple viewpoints yielded 684 hyperspectral images. Each image is comprised of 51 bands for those wavelengths ranging from 550 to 1,000 nm over 9 nm steps. For purposes of photometric calibration, we have also captured an image of a white Spectral on calibration target so as to recover the power spectrum of the illuminant across the scene. In Fig. 5, we show sample images in our dataset for the five categories of toys and the coins. In the figure, each column corresponds to one of these categories.

In our experiments, our descriptors are used for recognition as follows. We first partition the imagery into two sets of equal size. One set is used for purposes of training, whereas the other is used as a testing data-base for purposes of recognition.

Fig. 4 Hyperspectral images of multiple viewpoints of an object in the fluffy toy category in our dataset. The toy sample was acquired over ten views by rotating the object in increments of 10° about its vertical axis

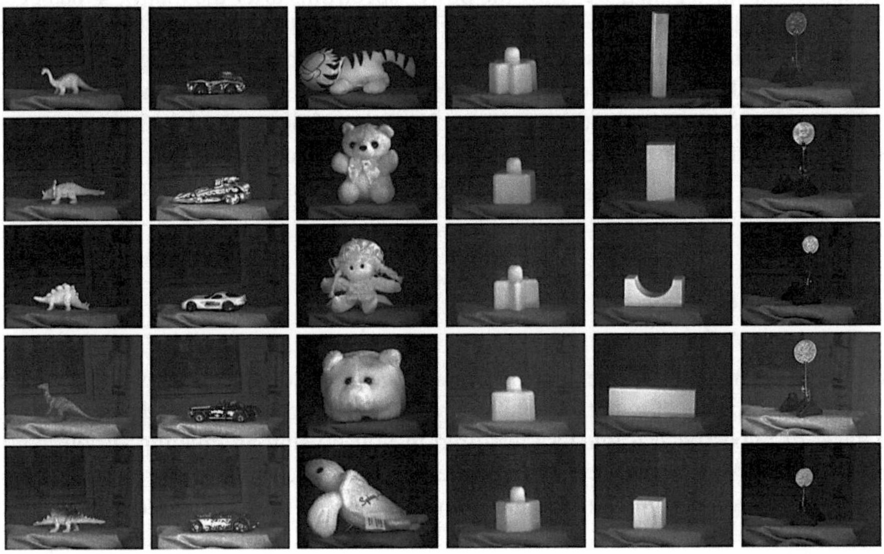

Fig. 5 Hyperspectral images of sample objects in each of the six categories in our dataset. These categories are, from *left-to-right*, plastic dinosaurs and animals, miniature cars, fluffy dolls, plastic blocks, wooden blocks and coins

The recognition task is performed by a k-nearest neighbour classifier [7] and a Support Vector Machine (SVM) [8]. For the SVM, we use a RBF kernel whose parameters have been obtained via cross validation.

Note that, to our knowledge, there is no hyperspectral image descriptors available in the literature. Nonetheless, it is worth noting that the wavelength resolved nature of hyperspectral imagery are reminiscent of the time dependency in dynamic textures, where a pixel in the image can be viewed as a stationary time series. As a result, we compare our results with those yielded using the algorithm in [43]. The reasons for this are twofold. Firstly, this is a dynamic texture descriptor based upon local

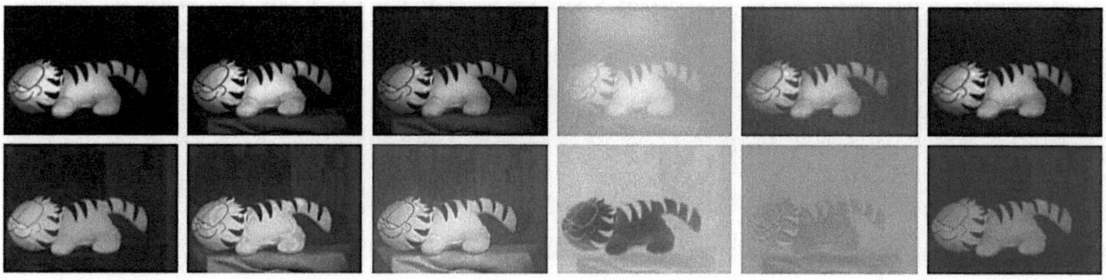

Fig. 6 From *left-to-right*: sample hyperspectral images of a fluffy toy at a number of wavelength-resolved bands, i.e., $\lambda = \{550, 640, 730, 820, 910, 1{,}000\,\text{nm}\}$. The *top row* shows the bands corresponding to the uncalibrated images and the *bottom row* shows the calibrated bands

binary patterns (LBPs), which can be viewed as a local definition of texture and shape in the scene which combines the statistical and structural models of texture analysis. Secondly, from the results reported in [43], this method provides a margin of advantage over other alternatives in the dynamic texture literature. For the descriptors, in the case of the LBP method in [43], we have used a dimensionality of 1938 over the 51 bands in the images. For our descriptor, the dimensionality is 1,500.

Since we have photometric calibration data available, in our experiments we have used two sets of imagery. The first of these corresponds to the dataset whose object images are given by the raw imagery. The second of these is given by the images which have been normalised with respect to the illuminant power spectrum. Thus, the first set of images corresponds to those hyperspectral data where the classification task is effected upon scene radiance, whereas the latter corresponds to a set of reflectance images. From now on, we denote the radiance-based set as the "uncalibrated" one, and the reflectance imagery as "calibrated". In Fig. 6, we show sample hyperspectral image bands for a fluffy toy at wavelengths corresponding to 550, 640, 730, 820, 910, and 1,000 nm. In the figure, the top row shows the uncalibrated imagery whereas the bottom row shows the calibrated data.

For purposes of recognition, we have computed our descriptors and the alternative making use of an approach reminiscent of the level-1 spatial histogram representation in [21]. This is, we have subdivided the images in a lattice-like fashion into 4, 16 and 64 squared patches of uniform size. In Fig. 7 we show the 4, 16 and 32-square lattice on the fluffy toy image. As a result, each image in either set, i.e., calibrated or uncalibrated, is comprised by 4, 16 or 64 descriptors. Here, we perform recognition based upon a majority voting scheme, where each of these descriptors is classified at testing time. Further, note that the fact that we have divided each image into 4, 16 and 64 squared regions provides a means to multiscale descriptor classification. Thus, in our experiments, we have used two majority voting schemes. The first of these limits the classification of descriptors to those at the same scale, i.e., number of squared regions in the image. The second scheme employs all the descriptors computed from multiple scales, i.e., $64 + 16 + 4$ for every image.

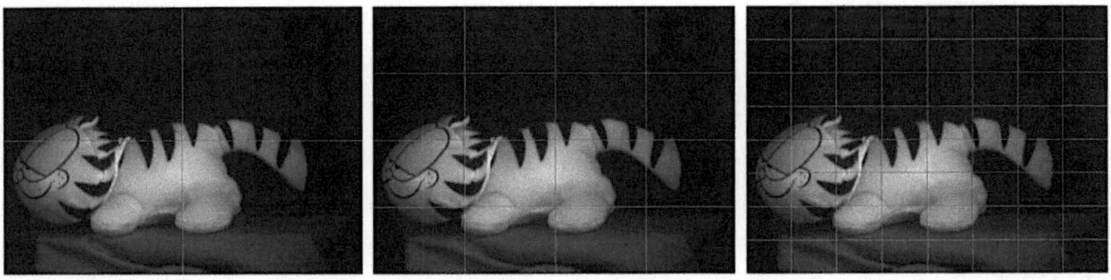

Fig. 7 From *left-to-right*: 4, 16 and 64-squared image region partitioning of the fluffy toy image

Table 1 Image categorisation results as percentage of correctly classified items in the dataset using the nearest neighbour classifier and our descriptor

Level	Category	Same scale (%)		Multiple scale (%)	
		Calibrated	Uncalibrated	Calibrated	Uncalibrated
4-Region lattice	Animals	97.39	90.32	99.13	99.13
	Cars	70.00	77.55	100.0	100.0
	Fluffy dolls	83.33	41.49	90.00	96.67
	Plastic blocks	80.00	96.24	97.14	97.14
	Wooden blocks	96.00	98.74	99.00	99.00
	Coins	93.75	87.64	96.88	96.88
	Average total	91.23	89.47	97.72	98.54
16-Region lattice	Animals	94.78	98.26	100.0	100.0
	Cars	90.00	80.00	100.0	100.0
	Fluffy dolls	80.00	93.33	96.67	96.67
	Plastic blocks	97.14	94.29	100.0	97.14
	Wooden blocks	100.0	100.0	99.00	99.00
	Coins	90.63	93.75	96.88	96.88
	Average total	94.44	95.91	99.12	98.83
64-Region lattice	Animals	98.26	98.26	97.39	97.39
	Cars	96.67	96.67	96.67	100.0
	Fluffy dolls	80.00	76.67	90.00	100.0
	Plastic blocks	82.86	82.86	97.14	94.29
	Wooden blocks	100.0	100.0	100.0	100.0
	Coins	90.63	90.63	100.0	96.88
	Average total	94.74	94.44	97.66	98.25
Average		93.47	93.27	98.17	98.54

In Tables 1, 2, 3 and 4 we show the categorisation results for our dataset. In the tables, we show the results, per category and overall average, for the calibrated and uncalibrated data for both classifiers over the two schemes described above, i.e., multiscale and single-scale, when both, our method and the alternative are used to compute the image descriptors for the imagery. From the tables, its clear that our descriptor delivers better categorisation performance consistently for both classifiers This is important since our descriptor has a lower dimensionality than the alternative. We can attribute this behaviour to the high information compaction of the FCT.

Table 2 Image categorisation results as percentage of correctly classified items in the dataset using a nearest neighbour classifier and the LBP-based descriptor in [43]

Level	Category	Single scale (%)		Multiple scale (%)	
		Calibrated	Uncalibrated	Calibrated	Uncalibrated
4-Region lattice	Animals	97.39	100.0	97.39	100.0
	Cars	30.00	93.33	6.67	93.33
	Fluffy dolls	88.57	97.14	80.00	97.14
	Plastic blocks	56.67	100.0	53.33	100.0
	Wooden blocks	52.00	98.00	40.00	98.00
	Coins	65.63	96.88	31.25	96.88
	Average total	65.04	97.56	51.44	97.56
16-Region lattice	Animals	94.78	99.13	96.52	96.52
	Cars	16.67	56.67	3.33	80.00
	Fluffy dolls	68.57	94.29	62.86	88.57
	Plastic blocks	13.33	70.00	20.00	13.33
	Wooden blocks	54.00	100.0	30.00	94.00
	Coins	18.75	90.63	3.13	6.25
	Average total	44.35	85.12	35.97	63.11
64-Region lattice	Animals	97.39	100.0	94.78	92.17
	Cars	0.00	0.00	0.00	3.33
	Fluffy dolls	45.71	54.29	51.43	65.71
	Plastic blocks	0.00	13.33	0.00	0.00
	Wooden blocks	33.00	98.00	28.00	93.00
	Coins	0.00	0.00	0.00	0.00
	Average total	29.35	44.27	29.04	42.37
Average		46.25	75.65	38.82	67.68

Also, note that for the nearest neighbour classifier, the overall results yielded using our method show no clear trend with respect to the use of reflectance, i.e., calibrated data, or radiance (uncalibrated imagery). This suggests that our method is robust to illuminant power spectrum variations. In the case of the SVM, the calibrated data with a multiscale approach delivers the best average categorisation results. For the alternative, the nearest neighbour classifier on uncalibrated data yields the best performance. Nonetheless, in average, absolute bests between the two descriptor choices here are 23% apart, being 75.63% for the LBP descriptor and 98.54% for our method. Further, note that for the coins, the alternative can be greatly affected by the effect of specularities at finer scales, i.e., the 64-region lattice. In contrast, our descriptor appears to be devoid of this sort of corruption.

Table 3 Image categorisation results as percentage of correctly classified items in the dataset using and SVM with an RBF kernel and our descriptor

Level	Category	Single scale (%)		Multiple scale (%)	
		Calibrated	Uncalibrated	Calibrated	Uncalibrated
4-Region lattice	Animals	97.39	97.39	97.39	99.13
	Cars	90.00	100.0	86.67	0.00
	Fluffy dolls	90.00	88.57	100.0	85.71
	Plastic blocks	100.0	100.0	97.14	96.67
	Wooden blocks	99.00	98.00	99.00	99.00
	Coins	100.0	96.88	78.13	96.88
	Average total	96.07	96.81	93.06	79.56
16-Region lattice	Animals	89.57	100.0	91.30	100.0
	Cars	100.0	70.00	96.67	0.00
	Fluffy dolls	63.33	62.86	100.0	22.86
	Plastic blocks	91.43	100.0	91.43	76.67
	Wooden blocks	100.0	100.0	99.00	94.00
	Coins	100.0	100.0	71.88	81.25
	Average total	90.72	88.81	91.67	62.46
64-Region lattice	Animals	90.43	100.0	94.78	33.04
	Cars	80.00	0.00	93.33	26.67
	Fluffy dolls	76.67	14.29	90.00	11.43
	Plastic blocks	56.67	0.00	82.86	26.67
	Wooden blocks	100.0	70.00	92.00	69.00
	Coins	53.13	96.88	78.13	90.63
	Average total	76.15	46.86	88.52	42.91
Average		87.65	77.49	92.37	47.48

8 Conclusion

In this chapter, we have shown how a local hyperspectral image descriptor can be generated via harmonic analysis. This descriptor is invariant to affine transformations on the corresponding local planar object surface patch. The descriptor is computed using an integral transform whose kernel is harmonic in nature. Affine invariance is then attained by relating the local planar object surface patch to a plane of reference whose orientation is fixed with respect to the camera plane. We have shown how high information compaction in the classifier can be achieved by making use of an FCT. It is worth stressing that the developments in the chapter are quite general and can be applied to a number of harmonic kernels spanning a Hilbert space. This opens-up the possibility of using other techniques available elsewhere in the literature, such as Mellin transforms, wavelets or Hankel transforms. We have shown the utility of the descriptor for purposes of image categorisation on a dataset of real-world hyperspectral images.

Table 4 Image categorisation results as percentage of correctly classified items in the dataset using and SVM with an RBF kernel and the LBP descriptor in [43]

Level	Category	Single scale (%)		Multiple scale (%)	
		Calibrated	Uncalibrated	Calibrated	Uncalibrated
4-Region lattice	Animals	93.91	98.26	80.87	100.0
	Cars	80.00	96.67	20.00	53.33
	Fluffy dolls	100.0	100.0	80.00	91.43
	Plastic blocks	70.00	100.0	3.33	66.67
	Wooden blocks	83.00	100.0	80.00	97.00
	Coins	93.75	100.0	6.25	100.00
	Average total	86.78	99.15	45.08	84.74
16-Region lattice	Animals	83.48	99.13	82.61	99.13
	Cars	31.03	44.83	0.00	3.45
	Fluffy dolls	65.71	80.00	42.86	51.43
	Plastic blocks	6.67	70.00	0.00	20.00
	Wooden blocks	70.00	99.00	70.00	98.00
	Coins	28.13	84.38	3.13	93.75
	Average total	47.50	79.56	33.10	60.96
64-Region lattice	Animals	79.35	83.48	77.39	88.70
	Cars	0.00	0.00	0.00	0.00
	Fluffy dolls	19.29	17.14	2.86	5.71
	Plastic blocks	0.00	0.00	0.00	0.00
	Wooden blocks	61.25	84.00	60.00	67.00
	Coins	0.78	0.00	3.13	3.13
	Average total	26.78	30.77	23.90	27.42
Average		53.69	69.83	34.03	57.72

References

1. Aloimonos, J., Swain, M.J.: Shape from patterns: regularization. Int. J. Comput. Vis. **2**, 171–187 (1988)
2. Asmussen, S., Binswanger, K., Hojgaard, B.: Rare events simulation for heavy-tailed distributions. Bernoulli **6**(2), 303–322 (2000)
3. Bar-Joseph, Z., El-Yaniv, R., Lischinski, D., Werman, M.: Texture mixing and texture movie synthesis using statistical learning. IEEE Trans. Visual. Comput. Graph. **7**(2), 120–135 (2001)
4. Bracewell, R.N., Chang, K.-.Y., Jha, A.K., Wang, Y.-.H.: Affine theorem for two-dimensional Fourier transform. Electron. Lett. **29**, 304–309 (1993)
5. Chen, Y., Wang, J.Z., Krovetz, R.: CLUE: cluster-based retrieval of images by unsupervised learning. IEEE Trans. Image Process. **14**(8), 1187–1201 (2005)
6. Chum, O., Philbin, J., Sivic, J., Isard, M., Zisserman, A.: Total recall: automatic query expansion with a generative feature model for object retrieval. In: Proceedings of the IEEE 11th International Conference on Computer Vision, 2007, ICCV 2007, pp. 1–8 (2007)
7. Cover, T., Hart, P.: Nearest neighbor pattern classification. IEEE Trans. Inf. Theory **13**(1), 21–27 (1967)
8. Cristianini, N., Shawe-Taylor, J., Elisseeff, A., Kandola, J.: On kernel-target alignment. Adv. Neural Inf. Process. Syst. **14**, 367–373 (2001)

9. Doretto, G., Chiuso, A., Wu, Y.N., Soatto, S.: Dynamic textures. Int. J. Comput. Vis. **51**(2), 91–109 (2003)
10. Dundar, M.M., Landgrebe, D.A.: Toward an optimal supervised classifier for the analysis of hyperspectral data. IEEE Trans. Geosci. Rem. Sens. **42**(1), 271–277 (2004)
11. Fazekas, S., Chetverikov, D.: Normal versus complete flow in dynamic texture recognition: a comparative study. In: Proceedings of the Fourth International Workshop Texture Analysis and Synthesis, pp. 37–42 (2005)
12. Fukunaga, K.: Introduction to Statistical Pattern Recognition, 2nd edn. Academic Press, New York (1990)
13. Gårding, J.: Direct estimation of shape from texture. IEEE Trans. Pattern Anal. Mach. Intell. **15**(11), 1202–1208 (1993)
14. Ghanem, B., Ahuja, N.: Phase based modelling of dynamic textures. In: Proceedings of the IEEE 11th International Conference on Computer Vision, ICCV 2007, pp. 1–8 (2007)
15. Ikeuchi, K.: Shape from regular patterns. Artif. Intell. **22**(1), 49–75 (1984)
16. Jimenez, L.O., Landgrebe, D.A.: Hyperspectral data analysis and feature reduction via projection pursuit. IEEE Trans. Geosci. Rem. Sens. **37**(6), 2653–2667 (1999)
17. Jolliffe, I.T.: Principal Component Analysis, 2nd edn. Springer, New York (2002)
18. Kanatani, K., Chou, T.: Shape from texture: general principle. Artif. Intell. **38**(1), 1–48 (1989)
19. Katznelson, Y.: An Introduction to Harmonic Analysis, 3rd edn. Cambridge University Press, Cambridge (2004)
20. Landgrebe, D.: Hyperspectral image data analysis. Signal Process. Mag. **19**(1), 17–28 (2002)
21. Lazebnik, S., Schmid, C., Ponce, J.: Beyond bags of features: spatial pyramid matching for recognizing natural scene categories. In: Proceedings of the 2006 IEEE Computer Society Conference on Computer Vision and Pattern Recognition, vol. 2, pp. 2169–2178 (2006)
22. Li, F.-F., Perona, P.: A Bayesian hierarchical model for learning natural scene categories. In: Proceedings of the 2005 IEEE Computer Society Conference on Computer Vision and Pattern Recognition, vol. 2, pp. 524–531 (2005)
23. Malik, J., Rosenholtz, R.: Computing local surface orientation and shape from texture for curved surfaces. Int. J. Comput. Vis. **23**(2), 149–168 (1997)
24. Nilsback, M., Zisserman, A.: A visual vocabulary for flower classification. In: Proceedings of the 2006. IEEE Computer Society Conference on Computer Vision and Pattern Recognition, vol. 2, pp. 1447–1454 (2006)
25. Nistér, D., Stewénius, H.: Scalable recognition with a vocabulary tree. In: Proceedings of the 2006 IEEE Computer Society Conference on Computer Vision and Pattern Recognition, vol. 2, pp. 2161–2168 (2006)
26. Otsuka, K., Horikoshi, T., Suzuki, S., Fujii, M.: Feature extraction of temporal texture based on spatiotemporal motion trajectory. In: Proceedings of the International Conference on Pattern Recognition, vol. 2, p. 1047 (1988)
27. Péteri, R., Chetverikov, D.: A brief survey of dynamic texture description and recognition. In: Computer Recognition Systems, pp. 17–26 (2005)
28. Portilla, J., Simoncelli, E.P.: A parametric texture model based on joint statistics of complex wavelet coefficients. Int. J. Comput. Vis. **40**(1), 49–71 (2000)
29. Quelhas, P., Monay, F., Odobez, J., Gatica-Perez, D., Tuytelaars, T., Van Gool, L.: Modelling scenes with local descriptors and latent aspects. In: Proceedings of the 10th IEEE International Conference on Computer Vision, pp. 883–890 (2005)
30. Rahman, A., Murshed, M.: A robust optical flow estimation algorithm for temporal textures. In: Proceedings of the 2005 International Conference on Information Technology: Coding and Computing (ITCC'05), vol. 2, pp. 72–76 (2005)
31. Randen, T., Husoy, J.H.: Filtering for texture classification: a comparative study. IEEE Trans. Pattern Anal. Mach. Intell. **21**(4), 291–310 (1999)
32. Rao, K.R., Yip, P.: Discrete Cosine Transform: Algorithms, Advantages, Applications. Academic Press Professional, Inc., San Diego (1990)

33. Ribeiro, E., Hancock, E.R.: Shape from periodic texture using the eigenvectors of local affine distortion. IEEE Trans. Pattern Anal. Mach. Intell. **23**(12), 1459–1465 (2001)
34. Sengupta, K., Boyer, K.L.: Using geometric hashing with information theoretic clustering for fast recognition from a large CAD modelbase. In: Proceedings of the IEEE International Symposium on Computer Vision, pp. 151–156 (1995)
35. Sheikh, Y., Haering, N., Shah, M.: Shape from dynamic texture for planes. In: Proceedings of the 2006 IEEE Computer Society Conference on Computer Vision and Pattern Recognition, vol. 2, pp. 2285–2292 (2006)
36. Shokoufandeh, A., Dickinson, S.J., Siddiqi, K., Zucker, S.W.: Indexing using a spectral encoding of topological structure. In: Proceedings of the IEEE Computer Society Conference on Computer Vision and Pattern Recognition, 1999, vol. 2, pp. 491–497 (1999)
37. Sivic, J., Zisserman, A.: Video Google: text retrieval approach to object matching in videos. In: Proceedings of the International Conference on Computer Vision, vol. 2, pp. 1470–1477 (2003)
38. Sneddon, I.N.: Fourier Transforms. Dover, New York (1995)
39. Stein, E.M., Weiss, G.: Introduction to Fourier Analysis on Euclidean Spaces. Princeton University Press, Princeton (1971)
40. Super, B.J., Bovik, A.C.: Shape from texture using local spectral moments. IEEE Trans. Pattern Anal. Mach. Intell. **17**(4), 333–343 (1995)
41. Varma, M., Ray, D.: Learning the discriminative power-invariance trade-off. In: Proceedings of the International Conference in Computer Vision (ICCV2007), pp. 1–8 (2007)
42. Young, N.: An Introduction to Hilbert Space. Cambridge University Press, Cambridge (1988)
43. Zhao, G., Pietikäinen, M.: Dynamic texture recognition using local binary patterns with an application to facial expressions. IEEE Trans. Pattern Anal. Mach. Intell. **29**(6), 915–928 (2007)

Tracking and Identification via Object Reflectance Using a Hyperspectral Video Camera

Hien Van Nguyen, Amit Banerjee, Philippe Burlina, Joshua Broadwater and Rama Chellappa

Abstract Recent advances in electronics and sensor design have enabled the development of a hyperspectral video camera that can capture hyperspectral datacubes at near video rates. The sensor offers the potential for novel and robust methods for surveillance by combining methods from computer vision and hyperspectral image analysis. Here, we focus on the problem of tracking objects through challenging conditions, such as rapid illumination and pose changes, occlusions, and in the presence of confusers. A new framework that incorporates radiative transfer theory to estimate object reflectance and particle filters to simultaneously track and identify an object based on its reflectance spectra is proposed. By exploiting high-resolution spectral features in the visible and near-infrared regimes, the framework is able to track objects that appear featureless to the human eye. For example, we demonstrate that near-IR spectra of human skin can also be used to distinguish different people in a video sequence. These capabilities are illustrated using experiments conducted on real hyperspectral video data.

Keywords Hyperspectral · Video tracking · ID · Reflectance · Particle filter

H. V. Nguyen (✉) · R. Chellappa,
Center for Automation Research, University of Maryland at College Park,
College Park, USA
e-mail: hien@umd.edu

R. Chellappa
e-mail: Rama@cfar.umd.edu

A. Banerjee · P. Burlina · J. Broadwater
Applied Physics Laboratory, Johns Hopkins University, Baltimore, USA

A. Banerjee
e-mail: Amit.Banerjee@jhuapl.edu

P. Burlina
e-mail: Philippe.Burlina@jhuapl.edu

J. Broadwater
e-mail: Joshua.Broadwater@jhuapl.edu

R. Hammoud et al. (eds.), *Machine Vision Beyond Visible Spectrum*,
Augmented Vision and Reality, 1, DOI: 10.1007/978-3-642-11568-4_9,

1 Introduction

Computer vision algorithms generally exploit shape and/or appearance features for automated analysis of images and video. Appearance-based methods are challenged by the changes in an object's appearance due to (a) different illumination sources and conditions, (b) variations of the object pose with respect to the illumination source and camera, and (c) different reflectance properties of the objects in the scene. Such variabilities compromise the performance of many vision systems, including background subtraction methods for motion detection, appearance-based trackers, face recognition algorithms, and change detection.

Most modern video cameras provide imagery with high spatial and temporal resolution that is ideal for detecting and tracking moving objects. However, their low spectral resolution limits their ability to classify or identify objects based on color alone [1, 5].

Conversely, hyperspectral sensors provide high spectral resolution imagery that the remote sensing community has utilized for a variety of applications, including mineral identification [11, 14] , land cover classification [2], vegetation studies [8], atmospheric studies [10] , search and rescue [18], and target detection [9, 17]. All of these applications depend on a material having a spectral "fingerprint" that can be gleaned from the data for measurement purposes. Using the idea of spectral "fingerprints", hyperspectral sensors can also be used for the identification and tracking of objects and people in ground-based applications. The problem until recently is that hyperspectral sensors did not provide a rapidly enough imaging of the environment to capture quickly moving objects such as people walking by in the near-field. In other words, hyperspectral sensors offer high spectral resolution imagery that is ideal for detection and identification. However, their low temporal resolution hinders its use for analyzing dynamic scenes with moving objects.

Recent groundbreaking work has led to the development of novel hyperspectral video (HSV) cameras that are able to capture hyperspectral datacubes at near video rates. This technological breakthrough was made possible by recent innovations in fast electronics and sensor design. While standard video cameras capture only three wide-bandwidth color images, the HSV camera collects many narrow-bandwidth images in the visible and near-infrared wavelengths. For surveillance systems, the HSV sensor provides the following advantages:

- High spatial and temporal resolution to detect and track moving objects.
- High spectral resolution to distinguish between objects with similar color.
- Ability to incorporate well-established radiative transfer theory models to mitigate the effects of illumination variations.

Since the hyperspectral video camera is able to simultaneously capture images with high temporal, spatial, *and* spectral resolution, it combines the advantages of both video and hyperspectral imagery in a single sensor package.

In this chapter, we develop a framework that combines hyperspectral video and particle filters to simultaneously track and ID objects in challenging conditions. The system demonstrates the following capabilities:

- *Illumination-insensitive tracking and ID.* Radiative transfer models are used to estimate the object's reflectance spectra, which are insensitive to a wide range of illumination variations.
- *Mitigating effects of confusers.* The high-resolution spectral features allow the system to distinguish the tracked object from others with similar color or appearance properties.
- *Non-continuous tracking.* By relying on the stable spectral features, the system is able to associate current observation of the tracked object with previous ones by utilizing spectral matching algorithms. Hence, the object can be tracked through spatial and temporal gaps in coverage.
- *Spectral "fingerprinting" of people.* The near-infrared wavelengths of the HSV camera provide spectral features to help distinguish the skin of tracked person from other persons in the scene.

A key component of the proposed system is the particle filter that jointly tracks and identifies the object. Instead of relying on a single image, it combines multiple observations to identify the tracked object in a principled Bayesian framework. We show how the confidence in the identification increases with additional observations.

This chapter is organized as follows. Section 2 briefly describes how the hyperspectral video camera captures hyperspectral images at high speeds. Section 3 details the approach to estimate illumination-invariant reflectance from the observed radiance measurements, and Sect. 4 discusses how the reflectance spectra are used to detect and track moving subjects in challenging conditions using a particle filter. An approach to identify the subjects from a sequence of observations is discussed in Sect. 5. Section 6 presents experiments on real hyperspectral video (HSV) data to demonstrate spectral-based tracking of objects through occlusions and illumination changes. Finally, Sect. 7 concludes the chapter.

2 The Hyperspectral Video Camera

Most modern video cameras provide imagery with high spatial and temporal resolution that is ideal for detecting and tracking moving objects. However, low spectral resolution limits their ability to classify or identify objects based on color alone. Conversely, traditional hyperspectral sensors offer high-resolution spectral and spatial imagery at low temporal resolutions with modest frame rates (up to 0.5 Hz). Hence, they have been utilized extensively for object detection and classification in static, non-dynamic environments.

The HSV camera is a passive sensor that measures the optical spectra of every pixel from 400 to 1,000 nm (visible and near-IR wavelengths). It acquires datacubes using a line scanning technique. An oscillating mirror scans the scene up to 10 times a

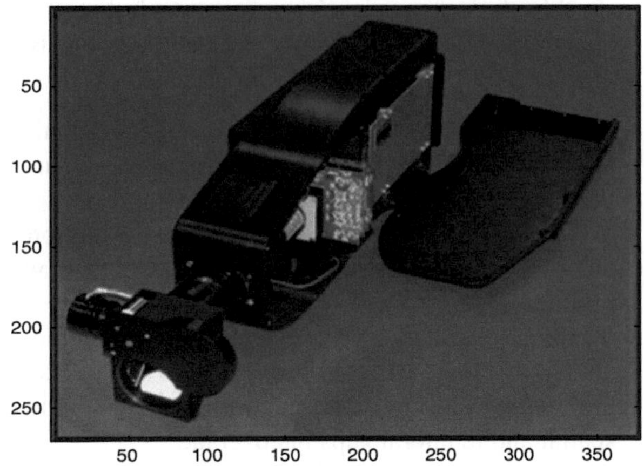

Fig. 1 The real-time HSV camera accepts traditional C-mount lenses (35 and 70 mm)

Table 1 HSV camera specifications

Camera weight	27 lbs.
Camera dimensions	$7'' \times 9.5'' \times 26''$
Spectral bandwidth	400–1,000 nm (visible and near-IR)
Number of bands	22, 44, 88
Maximum frame rate	10, 5, 2.5 cubes/s
CCD size	512×512 pixels
CCD pixel size	$18\,\text{mm} \times 18\,\text{mm}$
Dynamic range	16 bits/pixel

second, and for each mirror position one horizontal scan line is acquired and its pixels are decomposed by a spectrometer into a full spectral plane. The spectral plane is captured by a CCD, and is built into a datacube as the mirror completes a vertical scan of the scene. The acquired cube is then either immediately processed in real time or stored on a hard drive. Acquisition conditions are controlled by an operator through an on-board computer. Integration time can also be modified to accommodate low light conditions. The camera combines the benefits of video and hyperspectral data to simultaneously detect, track, and identify objects using well-established methods from computer vision and hyperspectral image analysis (Fig. 1).

The Surface Optics Corporation developed the HSV camera used in this study. Some of the important specifications of the camera are included in Table 1.

3 Estimating Object Reflectance

This section reviews the radiative transfer theory model that forms the basis for many hyperspectral image analysis algorithms. Given the HSV camera, we can *for the first time* apply this model to video tracking problems. Specifically, the high-resolution

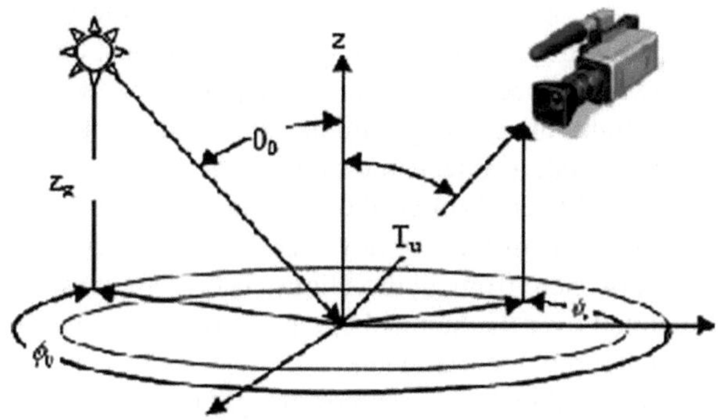

Fig. 2 Source–object–receiver geometry

spectra at each pixel can be analyzed to estimate its reflectance properties. The reflectance spectra are intrinsic to the objects and can be used to track and identify through a wide range of varying illumination and occlusions.

For this chapter, we are concerned only with the wavelengths of light from 400 to 1,000 nm (the visible to near infrared region of the electromagnetic spectrum). The algorithm described below applies principally for objects with Lambertian or diffuse surfaces. Highly specular objects will require additional processing which is beyond the scope of this chapter.

In the reflectance domain, there are six main sources of light [16]. The most obvious source is the sun. Light is generated at the sun, passes through the atmosphere, reflects off the object being imaged, and eventually reaches the sensor. Along the way, the spectral properties of the light are changed as photons are absorbed and scattered through the atmosphere. These effects can be modeled as

$$\begin{aligned} L_{\text{sun}}(x, y, \lambda) = {} & E_0(\lambda)\cos(\theta_0) T_d(z_g, \theta_0, \phi_0, \lambda) \\ & \times R(x, y, \lambda) T_u(z_g, z_u, \theta_0, \phi_v, \lambda) \end{aligned} \tag{1}$$

where L_{sun} is the light (or radiance) seen at the sensor generated from sunlight, E_0 is the exoatmospheric spectral signature of sunlight, T_d is the downward atmospheric transmittance, R is the reflectance of the object being imaged, and T_u is the upward atmospheric transmittance. All of these quantities are a function of the spectral wavelength λ and most of the quantities are based on the geometry of the source (sun), object being imaged, and receiver (camera) geometry as shown in Fig. 2. The geometries are based on cylindrical coordinates where z_g is the elevation of the sun, z_u is the elevation of the camera, θ_v is the declination of the camera from a normal vector to the surface, θ_0 is the declination of the sun from the same normal vector, ϕ_0 is the azimuth of the sun and ϕ_v is the azimuth of the camera.

A secondary source of light is skylight. Skylight can be modeled as

$$\begin{aligned} L_{\text{sky}}(x, y, \lambda) = {} & R(x, y, \lambda) T_u(z_g, z_u, \vartheta_v, \phi_v, \lambda) \\ & \times \int_{\phi=0}^{2\pi} \int_{\theta=0}^{\pi/2} E_s(\theta, \phi, \lambda)\cos(\vartheta)\sin(\theta)\, d\theta\, d\phi \end{aligned} \tag{2}$$

where L_{sky} is the skylight radiance seen at the sensor, R is the reflectance of the object being imaged, T_u is the upwelled atmospheric transmittance, and E_s is total downwelled spectral irradiation (i.e., the light energy scattered by the atmosphere).

Skylight takes a very similar path to sunlight. Once the light reaches the object being imaged, it reflects the same as the sunlight (assuming a diffuse material), and is reflected back through the atmosphere to the sensor along the same path as the sun light. The difference however is that skylight is generated by light scattered in the atmosphere from all directions. Therefore, these different patches of skylight are integrated over the hemisphere above the object being imaged which is represented by the double integral in (2).

The remaining four sources of light (upwelled radiance, multipath effects, adjacency effects, and trapping effects) are typically orders of magnitudes less energetic than sunlight or skylight. Because of this fact, these effects can largely be ignored for short-range, ground-based imaging. However, sometimes multipath and adjacency effects can become noticeable given the unique geometries of ground-based sensing. For example, light reflected from vegetation surrounding an object being imaged can impart part of the vegetative reflectance signature to the object—especially when the object is in full shade conditions where limited skylight is able to reach the object (e.g., dark shadows).

From the information above, the full radiometric equation is a simple sum of the six different sources of light [16]. Following the work of Piech and Walker [13], the radiometric transfer function for the model used in this research can be simplified to

$$L(x, y, \lambda) = R(x, y, \lambda)\,\{A(\lambda) + F(x, y)B(\lambda)\} \tag{3}$$

where $A(\lambda)$ represents the irradiance due to sunlight in (1), $F(x, y)$ represents the amount of sky light at pixel (x, y) (i.e., in shadow zones the amount of sky not blocked by the object creating the shadow), and $B(\lambda)$ represents the irradiance due to sky light in (2). Equation 3 assumes the scene is small enough that the source–object–receiver geometry is similar across the image and thus these terms are dropped. Also, the terms $A(\lambda)$ and $B(\lambda)$ are considered to be independent of pixel location when small areas are imaged (i.e., the sunlight and skylight terms do not vary over the small area being imaged).

Using the radiometric theory described, one can approximate the underlying reflectance signatures in an image. To do this, the image must contain an object with a known reflectance signature. One approach is to identify objects in the scene that have nearly flat reflectance signatures (i.e., constant and independent of wavelength) in full sunlight conditions. Examples of common materials with flat reflectance include concrete and asphalt. Using only those pixels that contain flat reflectances, (3) becomes

$$L_{\text{flat}}(\lambda) = kA(\lambda) + FB(\lambda) \tag{4}$$

where k now represents the unknown flat reflectance value independent of wavelength. One may also note that the location (x, y) has been removed as we only need to identify the flat reflectance object in the scene-not its location. If the entire image is in sunlight, then the reflectance of the image can be simply calculated as

$$\hat{R}(x, y, \lambda) = k\frac{L(x, y, \lambda)}{L_{\text{flat}}(\lambda)} \tag{5}$$

to within some unknown offset k. To remove the effects of k, each pixel is normalized to have the same energy. The result is an image with minimal illumination differences making tracking and identification of objects much simpler.

For images that contain shadow zones, the process is slightly more complicated. First, a shadow mask must be estimated. The energy of each pixel, computed using either the L_1 or L_2 norm of its spectra, is thresholded to produce the shadow mask. This algorithm is not perfect as very dark objects will be considered in shadow zones independent of their true illumination condition, but this approach produces acceptable results in an efficient manner.

Once a shadow mask has been created, a shadow line must be found that crosses across the same material. For the pixels in the full sunlight condition, (5) is applied to estimate the reflectance. Therefore, the reflectance of the material which is in both full sun and full shade conditions is found using (5). The resulting estimated reflectance $\hat{R}(x, y, \lambda)$ is used to calculate the skylight effects such that

$$kF(x, y)B(\lambda) = \frac{L_{\text{shadow}}(x, y, \lambda)}{\hat{R}(x, y, \lambda)} \tag{6}$$

where $L_{\text{shadow}}(x, y, \lambda)$ is the radiance of the same material in full shadow conditions. Now estimates for both full sun and full shade conditions are available.

Using these estimates and the shadow mask, pixels in full sun can be converted to reflectance using (5). For pixels in shade, their reflectance can be calculated using

$$R(x, y, \lambda) = \frac{L(x, y, \lambda)}{kF(x, y)B(\lambda)} \tag{7}$$

Again, we do not know the offsets due to k or $F(x, y)$, but this can be handled by normalizing the resulting reflectance as was done with (5).

The resulting image is an approximation to reflectance with most illumination variations removed as shown in Fig. 3. The first image in the figure is the color image derived from the original radiance spectra. The second image is the color image based on the estimated reflectance spectra for each pixel. The reflectance image has a number of desirable properties. First, the person on the right that was in full shadow now appears as if he is in full sunlight. Second, the bricks behind the bushes are now fully illuminated and the structures of the bushes can easily be seen. Third, most of the shadows in the grass have been removed making the grass appear fully illuminated. Finally, the concrete bench in the shadow region is now much more visible and its color/spectra is correctly estimated. In nearly all cases, the illumination variations throughout the image have been significantly reduced.

This does not happen in all cases however. One may note the shadows below the windows are still present. In these regions, the direct sunlight and skylight are

Fig. 3 Example of illumination insensitivity when using reflectance spectra. **a** The variations in *color* when using the radiance spectra are apparent. Computing the color from the reflectance spectra mitigates the effects of illumination variations, as seen in **(b)**

so weak, the multipath effects begin to dominate. The result is the bricks have a slight "green" tint to them from the light filtering through the trees and off the grass. This makes the estimated reflectance (which does not account for multipath effects) different from the surrounding bricks leaving a "shadow" area in the image. Nevertheless, tracking of objects in the approximate reflectance image can be easily accomplished as shown in Sect. 6.

4 Tracker Based on Particle Filtering

Particle filtering (PF) is an inference technique for estimating the unknown state θ_t from a noisy collection of observations $Y_{1...t} = Y_1, \ldots, Y_t$ arriving in sequential fashion. It was originally proposed in [3] in the signal processing literature. Since then, PF has been used to deal with many tracking problem that arises in computer vision [6]. We will lay out the general frame work of PF and explain how it is used for tracking in hyperspectral video sequences. We will also discuss how joint tracking and recognition is done in the PF framework.

4.1 Framework of Particle Filter

We define state vector $\theta_t = (m_t, n_t)$ which is also called a *particle*. m_t is motion vector whose parameters determine the transformation from an observation Y_t to an image patch Z_t, and n_t is the identity associated with that image patch. Intuitively, the knowledge of a state vector tells us the position and identity of a moving object. The set of N weighted particles at every time instant t decides the posterior probability $p(\theta_t | Y_{0:t})$ at that time instant, from which we estimate the true position of the tracked object.

4.1.1 Motion Equation

$$m_t = m_{t-1} + w_t \tag{8}$$

where w_t is *noise* in motion model, whose distribution determines the motion state transition probability. The choice of m_t is application dependent. In this study, we use the affine motion model which is good for the case where there is no significant pose variation available in the video sequence. More specifically,

$$m_t = \begin{pmatrix} a & b & t_x \\ c & d & t_y \end{pmatrix} \tag{9}$$

and coordinates of the patch of interest is computed as

$$\begin{pmatrix} x \\ y \end{pmatrix} = \begin{pmatrix} a & b \\ c & d \end{pmatrix} \begin{pmatrix} x_0 \\ y_0 \end{pmatrix} + \begin{pmatrix} t_x \\ t_y \end{pmatrix} \tag{10}$$

where $\begin{pmatrix} x_0 \\ y_0 \end{pmatrix}$ is the set of coordinates of template's pixels and $\begin{pmatrix} x \\ y \end{pmatrix}$ is the set of coordinates of a candidate's pixels. The first four parameters of motion vector $\begin{pmatrix} a & b \\ c & d \end{pmatrix}$ account for rotation and scaling effects while the last two parameters $\begin{pmatrix} t_x \\ t_y \end{pmatrix}$ are translation parameters.

4.1.2 Identity Equation

$$n_t = n_{t-1} \tag{11}$$

In this case, we assume the identity of a tracked object remains the same through time. In reality, a small transition probability among identities can be incorporated to account for the fact that the tracker sometimes confuses between different moving objects.

4.1.3 Observation Equation

By assuming that the transformed observation is a noise-corrupted version of a still template in the gallery, the observation equation can be written as

$$T_{m_t}(Z_t) = I_{n_t} + v_t \tag{12}$$

where v_t is observation noise at time t, whose distribution determines the observation likelihood $p(Z_t|m_t, n_t)$ and T_{m_t} is a transformed version of the observation Z_t.

We assume statistical independence between all noise variables and prior knowledge on the distributions $p(m_0|Z_0)$ and $p(n_0|Z_0)$. Recalling that the overall state vector is $\theta_t = (m_t, n_t)$, (8) and (11) can be combined into a single state equation which is completely described by the overall state transition probability

$$p(\theta_t|\theta_{t-1}) = p(m_t|m_{t-1})p(n_t|n_{t-1}) \tag{13}$$

For recognition, the goal is to compute the posterior probability $p(n_t|Z_0)$, which is in fact a marginal of the overall state posterior probability $p(\theta_t|Z_{0:t}) = p(m_t, n_t|Z_{0:t})$, where the marginalization is performed over the motion parameters. Similarly for tracking, the marginal probability $p(m_t|Z_{0:t})$ is required, which can be obtained by marginalizing the state posterior probability $p(\theta_t|Z_{0:t}) = p(m_t, n_t|Z_{0:t})$ over all identities n_t. These required posteriors can be computed using the particle-filtering framework based on the Sequential Importance Sampling (SIS) algorithm. We refer the reader to [19] for more details on how this is done.

4.2 Appearance Model

In order to illustrate the effectiveness of tracking reflectance images over radiance images, we use a fixed appearance model and do not incorporate the adaptive appearance model in this study. Intensity or spectral angle is utilized as criteria for comparison between a template and candidates. Obviously, this appearance model does not account for the variation in pose like in [20]. Nevertheless, we will show that the tracker can still perform well when videos are converted into reflectance.

4.2.1 Mean Square Intensity Error (MSIE)

$$\text{MSIE} = \sum_x (I(x) - \tilde{I}(x))^2 \tag{14}$$

where $x \in R^3$ is the spatial–spectral location in the hyperspectral data cube, I is the intensity of the template, and $\tilde{I}$ is the intensity of a candidate.

4.2.2 Mean Square Angle Error (MSAE)

$$\text{MSAE} = \sum_{x^*} \left(\frac{\langle S(x^*), \tilde{S}(x^*) \rangle}{|S(x^*)||\tilde{S}(x^*)|} \right)^2 \tag{15}$$

where $x^* \in R^2$ is the spatial location of the hyperspectral images, $S(x^*)$ is the spectral corresponding to location x^* of template, $\tilde{S}(x^*)$ is the spectral corresponding to location x^* of a candidate, and $\langle S(x^*), \tilde{S}(x^*) \rangle$ denotes the inner product of the two spectra.

4.3 Adaptive Velocity

With the availability of the sample set $M_{t-1} = \{m_{t-1}^{(j)}\}_{j=1}^{J}$ and the image patches of interest $Z_{t-1} = \{Z_{t-1}^{(j)}\}_{j=1}^{J}$ for a new observation Y_t, we can predict the shift in the motion vector (or adaptive velocity) $v_t = m_t - m_{t-1}$. The first-order linear approximation [4, 7] , which is essentially comes from the constant brightness constraint, i.e., there exists a m_t such that

$$T(Y_t; m_t) \approx \hat{Z}_{t-1} \tag{16}$$

Set $\tilde{m}_t = m_{t-1}$. Using a first-order Taylor series expansion around m_t yields

$$T(Y_t; m_t) \approx T(Y_t; \tilde{m}_t) + C_t(m_t - \tilde{m}_t) = T(Y_t; \tilde{m}_t) + C_t v_t \tag{17}$$

Combining (16) and (17) we have

$$\hat{Z}_{t-1} \approx T\{Y_t; \tilde{m}_t\} + C_t v_t \tag{18}$$

i.e.,

$$v_t = m_t - \tilde{m}_t \approx -B_t(T\{Y_t; \tilde{m}_t\} - \hat{Z}_{t-1}) \tag{19}$$

where B_t is the pseudo-inverse of the C_t matrix, which can be efficiently estimated from the available data M_{t-1} and Z_{t-1}.

5 Spectral Identification

Recent work has shown that the near-infrared spectra of human skin may be used to recognize people [12]. Studies suggest that a person's skin spectra is relatively stable across their face, while maintaining enough variability between subjects to aid in recognition. However, it has also been shown that the skin spectra of the same individual do vary with time and the condition of collection. For example, [12] showed that the top matching rate of 200 subjects based on their skin spectra reduced from around 90% to 60% over one week time due to the variation of skin spectra. Obviously, skin and hair spectra are not reliable to be used as permanent biometrics.

However, the spectra of many objects are often stable over a short time period and therefore can be exploited by hyperspectral imaging systems. For example, one can build a hyperspectral camera system capable of detecting a disguised robber seen within the last two days. Likewise, hyperspectral imaging enables the detection of customers wearing face masks to alert the security authority as soon as they enter into the place. These applications are possible only if we can show human related spectra such as those of hair and skin are good to be used as short-term biometrics [15].

5.1 Sequential Recognition

This section will show how we do spectral identification. Recall that under the PF framework, one can do recognition by marginalizing the posterior probability $p(\theta_t|Z_{0:t}) = p(m_t, n_t|Z_{0:t})$. However, to make the algorithm simple and applicable for all tracking frameworks, we choose tracking-then-recognition approach. Moreover, here only spectrals are used as primary source of information for matching. Let S_{n_t} be the spectral fingerprint associated with each identity n. For every observation at time instant t, a simple clustering algorithm is performed to eliminate outliers and retain only a set of spectrals $\mathbf{S}_t^{0:K} = \{S_t^1, S_t^2, \ldots, S_t^K\}$ resembling those spectral IDs in the database. The posterior is then computed for each identity as

$$P(n|\mathbf{S}_t^{0:K}) = \prod_{i=0}^{K} \frac{1}{\sqrt{2\pi|\Sigma|}} e^{-\frac{(S_t^i-\mu_t)^T\Sigma^{-1}(S_t^i-\mu_t)}{2}} \tag{20}$$

where μ_n is the mean spectrum of ID n, Σ is the spectral covariance matrix estimated from the data of all IDs. Assuming that observations $\mathbf{S}_{0:K}$ are independent through time. Subsequently, the cumulative probability for each identity is obtained simply by the multiplication of posterior probabilities of all time instants.

$$P(n|\mathbf{S}_{0:t}^{0:K}) = C \prod_{\tau=0}^{t} \prod_{i=0}^{K} \frac{1}{\sqrt{2\pi|\Sigma|}} e^{-\frac{(S_\tau^i-\mu_\tau)^T\Sigma^{-1}(S_\tau^i-\mu_t)}{2}} \tag{21}$$

where C is a normalization constant. The procedure decides on the identity n^* whenever the below condition is satisfied

$$P(n^*|\mathbf{S}_{0:t}^{0:K}) > \gamma P(n|\mathbf{S}_{0:t}^{0:K}), \quad \forall n \neq n^* \tag{22}$$

where γ is the threshold determined by empirical tests.

6 Experiments

We demonstrate the capabilities of the hyperspectral video tracking framework using three hyperspectral datasets. The first is the *tennis-shirts* sequence, which consists of 58 hyperspectral images captured at 5 Hz with 44 bands/cube. The video shows two subjects wearing red tennis shirts walking across the cameras field of view. They leave and re-enter the scene multiple times and walk in front and behind each other. Illumination is dramatically different along the pathway, since the right-hand side of the scene is under the shadow of a large tree. The challenges of this video come from small face regions, occlusions, and significant illumination changes (shadows). The second dataset is the *faces* sequence, which consists of 50 frames also captured at 5 Hz with 44 bands/cube. The video has five people moving from a shaded region

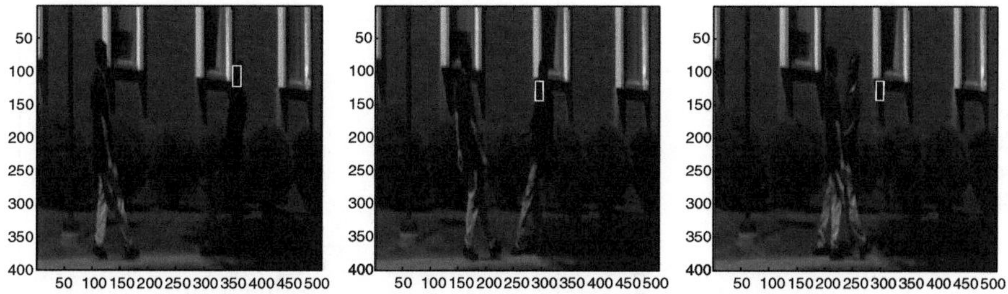

Fig. 4 Tracking an object based on its radiance spectra with a particle filter tracker

Fig. 5 Three frames from the *tennis-shirts* hyperspectral video tracking sequence. The two subjects are wearing *red tennis shirts* and walk in and out of shadow regions

toward the camera and eventually cutting to the right to leave the cameras field of view. Motion at the end of each sequence is very abrupt due to the close distance. Severe pose variation of turning faces at the end also makes this data set challenging. The *faces* sequence is used to evaluate the particle filter's ability to jointly track and identification the individuals in the scene. Finally, a collection of still hyperspectral images of 12 people's faces is used as the gallery for person recognition.

The radiance spectra for each of the hyperspectral datacubes are converted to reflectance using the method described in Sect. 3. The conversion to reflectance mitigates the effects of illumination changes and variations of the object's appearance. An example of the drawback of relying on the radiance spectra is shown in Fig. 4. Note how the tracker loses track of the dark object when its appearance changes due to sudden and brighter illumination changes.

Three frames from the *tennis-shirts* sequence are shown in Fig. 5; two subjects wearing red tennis-shirts that are nearly indistinguishable for the human eye or an RGB camera walk in and out of shadows. The objective is to use the spectra of the person's clothing to (a) track him through different illumination conditions, (b) distinguish him from the other person wearing a red tennis-shirt, and (c) maintain track while he walks in and out of the camera's field of view.

The difficulty of the problem is illustrated by comparing the scatter plots of the normalized color values (R, G)/(R + G + B) of the two red tennis-shirts in Fig. 6. The plots show that the color values of the two red shirts are not easily separable. They indicate that many appearance-based trackers using a traditional color camera would

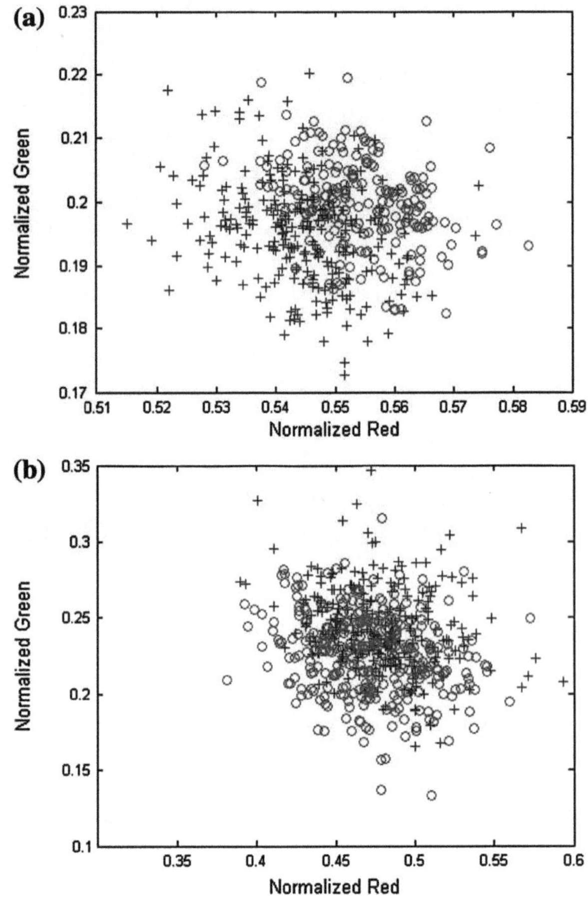

Fig. 6 Scatter plots of normalized *red* and *green* values of the two *red tennis-shirts*. Note how color does not provide separability between them in either illumination condition

Table 2 Posterior probabilities after five frames for ID 1 and ID 2

	Frame 1	Frame 2	Frame 3	Frame 4	Frame 5
(a) Correct identification of ID 1					
ID 1	*0.3437*	*0.521*	*0.6563*	*0.7555*	*0.8077*
ID 2	0.1704	0.1025	0.0601	0.0311	0.0191
ID 3	0.2647	0.2769	0.2449	0.1994	0.1683
ID 4	0.0943	0.0331	0.0106	0.003	0.0009
ID 5	0.1269	0.0664	0.0282	0.0111	0.004
(b) Correct identification of ID 2					
ID 1	0.2565	0.2523	0.2432	0.2138	0.1906
ID 2	*0.3017*	*0.442*	*0.5213*	*0.598*	*0.6635*
ID 3	0.2431	0.2188	0.1983	0.172	0.1393
ID 4	0.1034	0.0498	0.0224	0.0103	0.0043
ID 5	0.0953	0.0371	0.0149	0.0059	0.0022

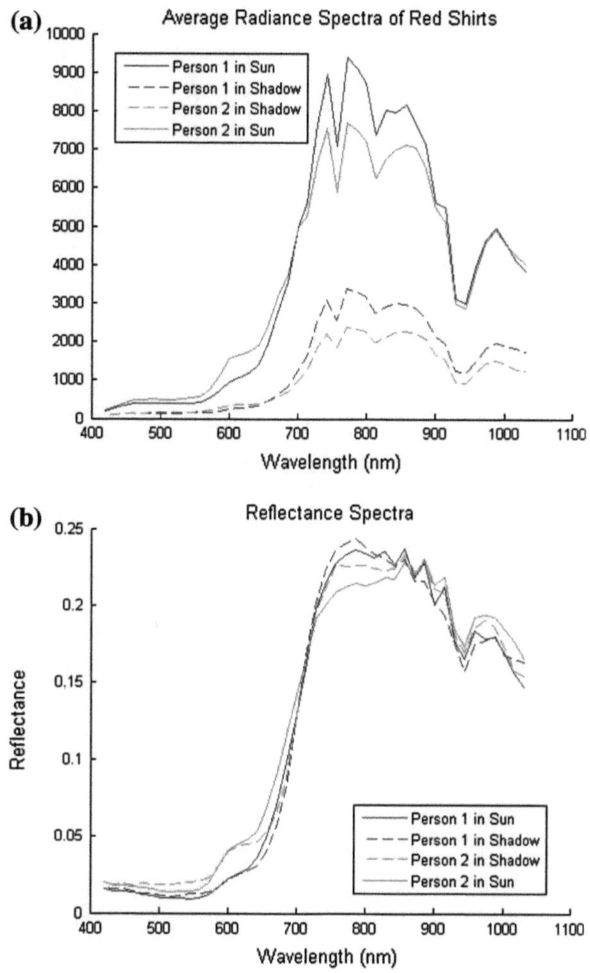

Fig. 7 Plots of the radiance and reflectance spectra of the *red tennis-shirts* in sun and shadow. Note how the reflectance spectra are consistent over different illumination conditions and can be used to distinguish between the two *red shirts*

be confused by the presence of both subjects, especially when they are in close proximity.

This problem is alleviated by the use of the full high-resolution spectra of the shirts. The plots of the observed radiance spectra of the red shirt are given in Fig. 7a. Note the spectral differences between the two shirts and how the spectra change when observed in sun and shadow. Using the algorithm described in Sect. 3, the reflectance spectra of the shirts are estimated from the observed radiance spectra. As can be seen in Fig. 7b, the estimated reflectance for each shirt is consistent, whether they are observed in sun or shadow. Further, the spectral differences in the two shirts are also maintained in the reflectance spectra. This indicates that the reflectance spectra can be used to track objects through varying illumination conditions and discriminate between the multiple objects with similar color.

Fig. 8 Skin tracking example. Note how the particle filter is able to track the person through shadows and in the presence of a confuser in the *tennis-shirts* sequence. By computing the reflectance spectrum for each pixel, the algorithm is able to track through varying illumination conditions and distinguish between the two people wearing *red tennis-shirts*. The use of the reflectance spectra also allows for the algorithm to maintain track of the object even when the person leaves and re-enters the scene

Two sample sequences of tracking human skin are shown in Figs. 8 and 10. Figure 9 shows a tracking sequence of red tennis-shirt in the presence of another confusing object with similar shirt spectra. Our algorithm successfully track all sequences where the traditional tracking using RGB features fails.

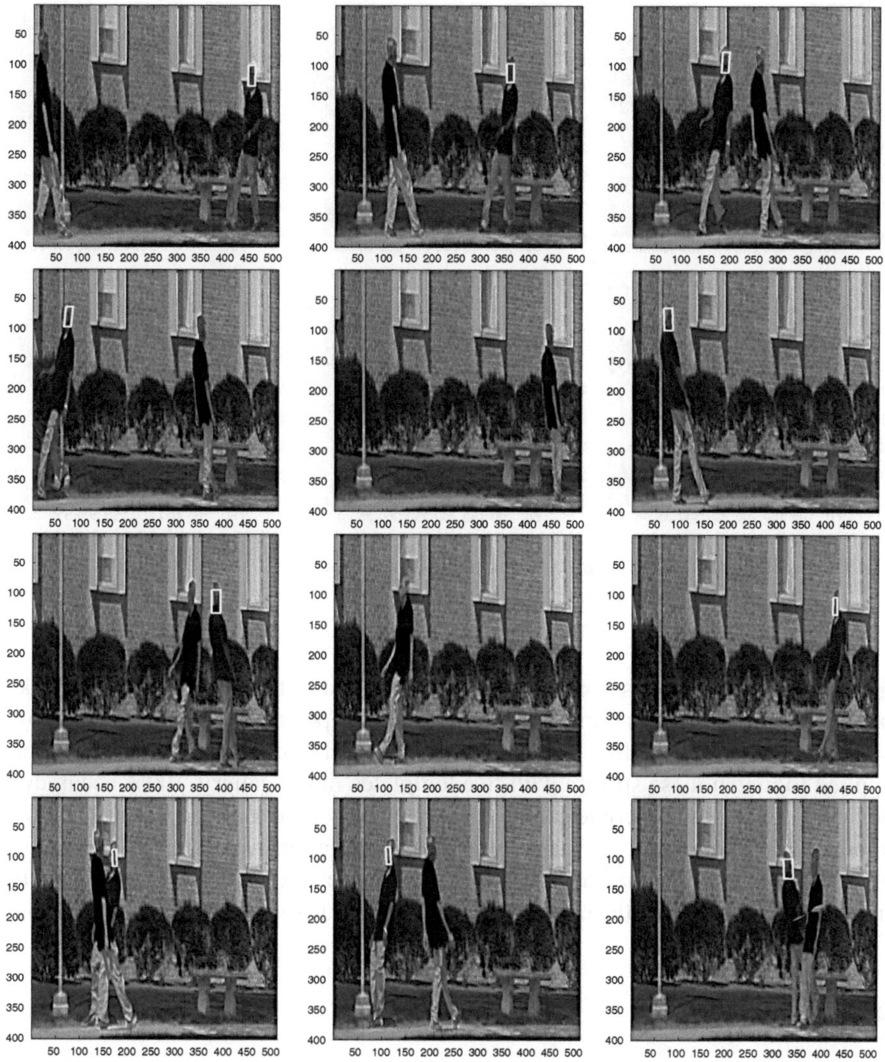

Fig. 9 Shirt tracking example. Note how the particle filter is able to track the person through shadows and in the presence of a confuser in the *red tennis-shirts* sequence. By computing the reflectance spectrum for each pixel, the algorithm is able to track through varying illumination conditions and distinguish between the two people wearing *red tennis-shirts*. The use of the reflectance spectra also allows for the algorithm to maintain track of the object even when the person leaves and re-enters the scene

Human identification: This experiment is designed to validate the assumption that it is possible to distinguish different people only by their skin spectra in a video sequence. Skin spectra of five people in the gallery are manually collected to be used as gallery. We make sure the frames in which these spectra are collected are not

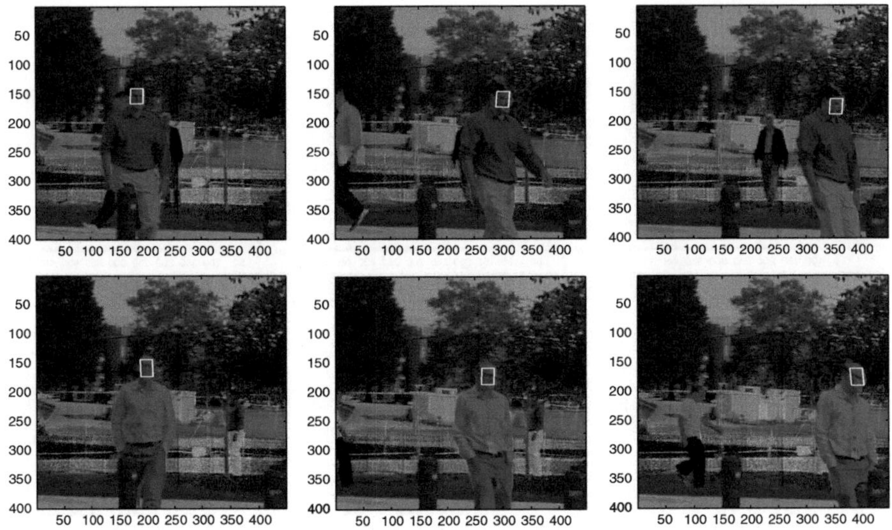

Fig. 10 Tracking faces with pose variations in the faces sequence

included in probe frames. First the tracker is applied to track faces of people. Skin pixels in cropped images corresponding to tracker's positions are then selected by a simple clustering algorithm. The recognition algorithm described in Sect. 5 is then applied to match these spectra with five templates in the gallery. 100% accuracy is achieved. Each of five people can be easily recognized only after five video frames. Table 2 shows how the posterior probabilities evolve after five iterations. It is noticeable in the second table that there is some confusion between ID 1, ID 2, and ID 3 in the first iteration. However, the confusion is completely resolved after five iterations.

7 Conclusion

In this chapter, we described an approach for detection and tracking of moving humans in a hyperspectral video sequence. Preliminary results on simultaneous tracking and recognition were also presented. The advantages of tracking and recognition using a hyperspectral video were demonstrated. In the immediate future, we will evaluate this approach on a large data set of humans.

Acknowledgments This research was supported by a Grant from JHU/Applied Physics Laboratory and the ONR MURI Grant N00014-08-1-0638.

References

1. Finlayson, G.D.: Computational color constancy. In: International Conference on Pattern Recognition, vol. 1, p. 1191 (2000)
2. Gianinetto, M., Lechi, G.: The development of superspectral approaches for the improvement of land cover classification. IEEE Trans. Geosci. Rem. Sens. **42**(11), 2670–2679 (2004)
3. Gordon, N.J., Salmond, D.J., Smith, A.F.M.: Novel approach to nonlinear/non-Gaussian Bayesian state estimation. IEE Proc. F Radar Signal Process. **140**(2), 107–113 (1993)
4. Hager, G.D., Belhumeur, P.N.: Efficient region tracking with parametric models of geometry and illumination. IEEE Trans. Pattern Anal. Mach. Intell. **20**(10), 1025–1039 (1998)
5. Healey, G., Slater, D.: Global color constancy: recognition of objects by use of illumination-invariant properties of color distributions. J. Opt. Soc. Am. A **11**(11), 3003–3010 (1994)
6. Isard, M., Blake, A.: CONDENSATION—conditional density propagation for visual tracking. Int. J. Comput. Vis. **29**(1), 5–28 (1998)
7. Jurie, F., Dhome, M.: A simple and efficient template matching algorithm. In: ICCV, pp. 544–549 (2001)
8. Lewis, M., Jooste, V., de Gasparis, A.A.: Discrimination of arid vegetation with airborne multispectral scanner hyperspectral imagery. IEEE Trans. Geosci. Rem. Sensing **39**(7), 1471–1479 (2001)
9. Manolakis, D.: Detection algorithms for hyperspectral imaging applications: a signal processing perspective, pp. 378–384 (2003)
10. Marion, R., Michel, R., Faye, C.: Measuring trace gases in plumes from hyperspectral remotely sensed data. IEEE Trans. Geosci. Remote Sens. **42**(4), 854–864 (2004)
11. Mustard, J.F., Pieters, C.M.: Photometric phase functions of common geological minerals and applications to quantitative analysis of mineral mixture reflectance spectra. J. Geophys. Res. **94**, 13619–13634 (1989)
12. Pan, Z., Healey, G., Prasad, M., Tromberg, B.: Face recognition in hyperspectral images. IEEE Trans. Pattern Anal. Mach. Intell. **25**(12), 1552–1560 (2003)
13. Piech, K.R., Walker, J.E.: Interpretation of soils. Photogrammetric Eng. **40**, 87–94 (1974)
14. Szeredi, T., Lefebvre, J., Neville, R.A., Staenz, K., Hauff, P.: Automatic endmember extraction from hyperspectral data for mineral exploration. In: 4th International Airborne Remote Sensing Conference Exhibition/21st Canadian Symposium on Remote Sensing, Ottawa, Ontario, Canada, pp. 891–896 (1999)
15. Satter, R.G.: 65 million in jewelry stolen from London store. http://abcnews.go.com/International/wireStory?id=8302495, Aug. 2009 (2009)
16. Schott, J.R.: Remote Sensing: The Image Chain Approach, 2nd edn. Oxford University Press, New York (2007)
17. Stein, D.W.J., Beaven, S.G., Hoff, L.E., Winter, E.M., Schaum, A.P., Stocker, A.D.: Anomaly detection from hyperspectral imagery. IEEE Signal Process. Mag. **19**(1), 58–69 (2002)
18. Subramanian, S., Gat, N.: Subpixel object detection using hyperspectral imaging for search and rescue operations. SPIE **3371**, 216–225 (1998)
19. Zhou, S.H., Krueger, V., Chellappa, R.: Probabilistic recognition of human faces from video. Comput. Vis. Image Understanding **91**(1–2), 214–245 (2003)
20. Zhou, S.H.K., Chellappa, R., Moghaddam, B.: Visual tracking and recognition using appearance-adaptive models in particle filters. IEEE Trans. Image Process. **13**(11), 1491–1506 (2004)

Moving Object Detection and Tracking in Forward Looking Infra-Red Aerial Imagery

Subhabrata Bhattacharya, Haroon Idrees, Imran Saleemi, Saad Ali and Mubarak Shah

Abstract This chapter discusses the challenges of automating surveillance and reconnaissance tasks for infra-red visual data obtained from aerial platforms. These problems have gained significant importance over the years, especially with the advent of lightweight and reliable imaging devices. Detection and tracking of objects of interest has traditionally been an area of interest in the computer vision literature. These tasks are rendered especially challenging in aerial sequences of infra red modality. The chapter gives an overview of these problems, and the associated limitations of some of the conventional techniques typically employed for these applications. We begin with a study of various image registration techniques that are required to eliminate motion induced by the motion of the aerial sensor. Next, we present a technique for detecting moving objects from the ego-motion compensated input sequence. Finally, we describe a methodology for tracking already detected objects using their motion history. We substantiate our claims with results on a wide range of aerial video sequences.

Keywords Aerial image registration · Object detection · Tracking

S. Bhattacharya (✉) · H. Idrees · I. Saleemi · M. Shah
University of Central Florida, 4000 Central Florida Blvd., Orlando, FL 32826, USA
e-mail: subh@cs.ucf.edu

H. Idrees
e-mail: haroon@cs.ucf.edu

I. Saleemi
e-mail: imran@cs.ucf.edu

M. Shah
e-mail: shah@cs.ucf.edu

S. Ali
Sarnoff Corporation, 201 Washington Road, Princeton, NJ 08540, USA
e-mail: sali@sarnoff.com

R. Hammoud et al. (eds.), *Machine Vision Beyond Visible Spectrum*, Augmented Vision and Reality, 1, DOI: 10.1007/978-3-642-11568-4_10,

1 Introduction

Detection and tracking of interesting objects has been a very important area of research in classical computer vision where objects are observed in various sensor modalities, including EO and IR, with static, hand-held and aerial platforms [34]. Many algorithms have been proposed in the past that differ in problem scenarios especially in camera dynamics and object dynamics [14, 15]. Tracking of a large, variable number of moving targets has been a challenging problem due to the sources of uncertainty in object locations, like, dynamic backgrounds, clutter and occlusions, and especially in the scenario of aerial platforms, measurement noise. In recent years, a significant amount of published literature has attempted to deal with these problems, and novel approaches like tracking-by-detection have been increasingly popular [17]. Such approaches involve the process of continuously applying a detection algorithm on single frames and associating detections across frames. Several recent multi-target tracking algorithms address the resulting data association problem by optimizing detection assignments over a large temporal window [2, 5, 17, 24].

Aerial tracking of multiple moving objects is however much more challenging because of the small object sizes, lack of resolution, and low quality imaging. Appearance based detection methods [10] are therefore, readily ruled out in such scenarios. The motion based object detection approaches rely on camera motion stabilization using parametric models [20], but in addition to parallax, cases of abrupt illumination changes, registration errors, and occlusions severely affect detection and tracking in airborne videos. Many algorithms have been proposed to overcome these problems of detection and tracking on frame to frame and pixel to pixel bases, including global illumination compensation [32], parallax filtering [37], and employing contextual information for detection [13, 29]. Some existing algorithms have performed well in planar scenes where adequate motion based foreground–background segmentations are achievable [36]. Most of the existing methods however have concentrated on medium and low altitude aerial platform sequences. Although such sequences suffer from the problem of strong parallax induced by structures perpendicular to the ground plane, like trees, towers, they do offer more pixels per target.

Effective use of visual data generated by UAVs requires design and development of algorithms and systems that can exhaustively explore, analyze, archive, index, and search this data in a meaningful way. In today's UAV video exploitation process, a ground station controls the on-board sensors and makes decisions about where the camera mounted on the bottom of the UAV should be looking. Video is relayed back to the intelligence center or some standard facility for assessment by the analysts. Analysts watch the video for targets of interest and important events which are communicated back to soldiers and commanders in the battle zone. Any post collection review of the video normally takes several hours for analysts to inspect a single video. The inherent inefficiency of this process and sheer magnitude of the data leads to an inability to process reconnaissance information as fast as it becomes available. The solution to this problem lies in augmenting the manual video exploitation process with computer vision based systems that can automatically manage and process ever increasing volume of aerial surveillance information without or with minimal involvement of human analyst. Such systems should handle all tasks from video reception to video registration, region of interest (ROI) detection to target tracking and event detection to video indexing. It should also be able to derive higher level semantic information from the videos which can be used to search and retrieve a variety of videos. Unfortunately, however there is still a gap between the operational requirements and the available capabilities in today's system for dealing with the UAV video stream.

A system capable of performing the above mentioned tasks for UAV videos will have to grapple with significantly higher levels of complexity as compared to the static camera scenario, as both the camera and the target objects are mobile in a dynamic environment. A significant amount of literature in the computer vision community has attempted to deal with some of these problems individually. We present a brief overview of these methods individually, along with the challenges and limitations involved.

1.1 Ego-Motion Compensation

Tracking of moving objects from a static camera platform is a relatively easier task than those from mobile platforms and is efficiently accomplished with sophisticated background subtraction algorithms. For a detailed study of these tracking techniques, the interested reader is requested to refer to [25]. Cameras mounted on mobile platforms, as observed in most aerial surveillance or reconnaissance, tend to capture unwanted vibrations induced by mechanical parts of the platform coupled with directed translation or rotation of the whole platform in 3-dimensional space. All the aforementioned forms of motion render even the most robust of the background subtraction algorithms ineffective in scenarios that involve tracking from aerial imagery.

A straightforward approach to overcome this problem is to eliminate the motion induced in the camera through the aerial platform, which is also known as *ego-motion compensation* in computer vision literature [12, 33, 35]. The efficacy of almost all image-based ego-motion compensation techniques depends on the underlying image registration algorithms they employ.

This step is also known as video alignment [9, 26] where objective is to determine the spatial displacement of pixels between two consecutive frames. The benefit of performing this step comes from the fact that after aligning the video, the intensity of only those pixels will be changing that correspond to moving objects on the ground. A detailed survey of various image alignment and registration techniques is available in [26]. Ideally an alignment algorithm should be insensitive to platform motion, image quality, terrain features and sensor modality. However, in practice these algorithms come across several problems:

- Large camera motion significantly reduces the overlap between consecutive frames which does not provide sufficient information to reliably compute the spatial transformation between the frames.
- Most of the alignment algorithms assume presence of dominant plane which is defined as a planar surface covering majority of pixels in an image. This assumption does not remain valid when a UAV views a non-planar terrain or takes a close up view of the object, which results in presence of multiple dominant planes. This causes parallax which often is hard to detect and remove.
- Sudden illumination changes result in drastic pixel intensity variations and make it difficult to establish feature correspondences across different frames. Gradient based methods for registration are more robust to an illumination change, rather than the feature based methods. Motion blur in the images can also throw off the alignment algorithm.

1.2 Regions of Interest Detection

Once the motion of the moving platform is compensated the next task is to identify 'regions of interest' (ROIs) from the video, the definition of which varies with application. In the domain of wide area surveillance employing UAVs, all the moving objects fall under the umbrella of ROI. Reliable detection of foreground regions in videos taken by UAVs poses a number of challenges, some of which are summarized below:

- UAVs often fly at a moderate to high altitude thus gathering the global context of the area under surveillance. Therefore, sizes of the potential target objects often appear very small in the range of 20–30 pixels. Small number of pixels on a target makes it difficult to distinguish it from the background and noise.
- As a UAV flies around the scene, the direction of illumination source (Sun) is continuously changing. If the background model is not constantly updated that may results in spurious foreground regions.
- Sometimes there are uninteresting moving objects present in the scene e.g., waving fags, flowing water, or moving leaves of a tree. If a background subtraction method falsely classifies such a region as a foreground region, then this region will be falsely processed as a potential target object.

1.3 Target Tracking

The goal of tracking is to track all the detected foreground regions as long as they remain visible in the field of view of the camera. The output of this module consists of trajectories that depict the motion of the target objects. In case of UAV videos several tracking options are available. One can perform tracking in a global mosaic or opt for tracking using geographical locations of the objects. Tracking in geographical locations is often called geo-spatial tracking and requires sensor modeling. Tracking algorithms also have to deal with number of challenges:

- Due to the unconstrained motion of the camera it is hard to impose constraints of constant size, shape, intensity, etc., on the tracked objects. An update mechanism needs to be incorporated to handle the dynamic changes in appearance, size and shape models.
- Occlusion is another factor that needs to be taken into account. Occlusions can be inter-object or caused by the terrain features e.g trees, buildings, bridges, etc.
- Restricted field of view of the camera adds to the complexity of the tracking problem. Detected objects are often geographically scattered. Restricted field of view of the camera allows UAV to track only certain number of objects at a time. It either has to move back and forth between all previously detected object or has to prioritize which target to pursue based upon the operational requirement.
- Tracking algorithms also have to deal with the imperfections of the object detection stage.

While designing a computer vision system that is capable of performing all the above mentioned tasks effectively in infra-red sequences, we need to consider the following additional issues:

- FLIR images are captured in significantly lower resolution compared to their EO counterparts as for a given resolution, infra-red sensor equipments are comparatively more expensive to install and maintain.
- FLIR sensing produces noisier images than regular EO imaging systems.
- As FLIR images tend to have lower contrast, they require further processing to improve the performance of algorithms used in ego-motion compensation, ROI detection and tracking.

The rest of this chapter is organized as follows: in Sect. 2 we discuss some of the prominent advances in the field of automatic target detection and tracking from aerial imagery. Section 3 provides a detailed description of our system that we have developed for tracking of objects in aerial EO/FLIR sequences. This section is followed by experimental results on 38 sequences from the VIVID-3 and AP-HILL datasets, obtained under permission from the Army Research Lab and US Govt.'s DARPA programs, respectively. We conclude the chapter with some of the limitations that we intend to address in future.

2 Related Work

Tracking moving objects from an aerial platform has seen numerous advances [1, 3, 16, 18, 30, 31, 38] in recent years. We confine our discussion to only a subset of the literature that has strong relevance with the context of this chapter. The authors of [16] present a framework that involves separating aerial videos into the static and dynamic scene components using 2-D/3-D frame-to-frame alignment followed by scene change detection. Initially, local tracks are generated for detected moving objects which are then converted to global tracks using geo-registration with a controlled reference imagery, elevation maps and site models. The framework is also capable of generating mosaics for enhanced visualization.

Zhang and Yuan [38] address the problem of tracking vehicles from a single moving airborne camera under occluded and congested circumstances using a tracker that is initialized from point features extracted from selected region of interest. In order to eliminate outliers that are introduced due to partial occlusion, an edge feature based voting scheme is used. In case of total occlusion, a Kalman predictor is employed. Finally, an appearance based matching technique is used to ensure that the tracker correctly re-associates objects on their re-entry into the field of view.

In [3], the authors use a video processor that has embedded firmware for object detection and feature extraction and site modeling. A multiple hypothesis tracker is then initialized using the positions, velocities and features to generate tracks of current moving objects along with their history.

The authors of [18] address the issue of urban traffic surveillance from an aerial platform employing a coarse-to-fine technique consisting of two stages. First, candidate regions of moving vehicle are obtained using sophisticated road detection algorithms followed by elimination of non-vehicle regions. In the next stage, candidate regions are refined using a cascade classifier that reduces the false alarm rate for vehicle detection.

Yalcin et al. [31] propose a Bayesian framework to model dense optical flow over time which is used to explicitly estimate the appearance of pixels corresponding to the background. A new frame is segregated into background and foreground object using an EM-based motion segmentation which is initialized by the background appearance model generated from previous frames. Vehicles on ground can be eventually segmented by building a mosaic of the background layer.

Xiao et al. [30] in their paper on moving vehicle and person tracking in aerial videos present a combination of motion layer segmentation with background stabilization, for efficient detection of objects. A hierarchy of gradient based vehicle versus person classifier is used on the detected objects prior to the generation of tracks.

The COCOA system [1] presented by Ali et al. is a 3-staged framework built using MATLAB, capable of performing motion compensation, moving object detection and tracking on aerial videos. Motion compensation is achieved using direct frame to frame registration which is followed by an object detection algorithm that relies on frame differencing and background modeling. Finally, moving blobs are tracked as long as the objects remain in the field of view of the aerial camera. The system has demonstrated its usability in both FLIR and EO scenarios.

The COCOALIGHT system is built from scratch keeping speed and portability into consideration while supporting the core functionalities of [1]. A detailed analysis of the algorithms employed for motion compensation, object detection and tracking with the justification behind their selection is provided in this chapter. We intend to disburse the technical insight while developing a practical system that is targeted to solve some of the predominant problems encountered while tracking in aerial imagery both within and beyond visible spectrum.

3 COCOALIGHT System Overview

The COCOALIGHT system shares the concept of modularity from its predecessor COCOA with complete change in design and implementation to facilitate tracking with near real-time latency. The software makes use of a widely popular open-source computer vision library which helps in seamlessly building the application both in 32 and 64-bit Windows and Linux PC platforms. Since the system is compiled natively, it is inherently much faster than interpreted MATLAB instructions present in COCOA. Furthermore, the software is packaged as an easy to use command-line console application eliminating memory intensive user interfaces from COCOA, rendering it a program with a low memory footprint, justifying the name COCOALIGHT. The design also exploits computational benefits from multi-threading during important fundamental image processing operations, e.g., gradient computation, feature extraction, computation of image pyramids.

Similar to COCOA, this system also consists of three independent components. However, unlike the COCOA system, which only supports processing in batch mode (an entire sequence needs to be processed to generate tracks), COCOALIGHT has capability for both batch and online processing. In the online mode, the tracking algorithm can be initialized with as few as only first ten frames from the video sequence. In addition, the software can leverage FFMPEG library support to process encoded videos without decompressing the video into image frames which is a significant improvement in usability over its MATLAB counterpart.

Having provided some knowledge about the implementation, we proceed towards a detailed discussion of the individual modules of the system.

3.1 Motion Compensation

The *motion compensation* module is the first and foremost module of the COCOALIGHT software framework. Any errors incurred in this module while eliminating camera motion get propagated to the subsequent modules namely the object detection and tracking modules. Due to this fact, the motion compensation stage necessitates employing highly accurate image alignment algorithms. Motivated solely by this objective we investigated several image alignment algorithms to suit our requirement. All our experiments are performed on sequences from VIVID dataset and from three other datasets, each collected using different aerial platforms flying over different geographical locations under different illumination conditions.

A study of the image registration techniques [9, 26] reveals that a registration algorithm must address the following issues which need careful consideration:

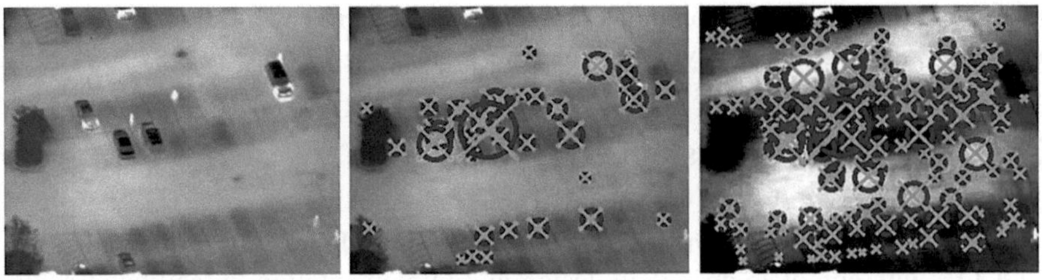

(a) (b) (c)

Fig. 1 Effect of histogram equalization on the detection of SURF interest points on a low contrast FLIR image. **a** Original FLIR image, **b** has a total of 43 SURF interest points whereas, **c** has a total number of 174 interest points after histogram equalization

- detecting candidate features also known as control points, from image pair to be registered,
- establishing correspondence between pairwise candidate features,
- estimating transformation model from point correspondence, and
- mapping image pair using the computed transformation model.

From our collection of video sequences, we observe that most of the frames demonstrate perspective projection artifacts. For this reason, we set our registration algorithm to estimate projective transformation parameters, also known as homography. Once homography parameters are obtained, a standard technique is available to perform the mapping operation between image pairs. In this section, we concentrate on the steps that involve proper selection of candidate features and establishing correspondence between the feature pairs.

In order to enhance feature detection in FLIR imagery, all the frames are subjected to a pre-processing stage. Histogram equalization is a widely popular technique to improve contrasts of IR images that are usually blurry. The effect of histogram equalization is clearly evident in the images shown in Fig. 1 with the histogram equalized image producing more interest points denoted by red–green circular cross-hairs.

3.1.1 Gradient-Based Method

Featureless spatio-temporal gradient-based methods are widely popular in image registration literature [9, 26] because of their ease of implementation. We use the unweighted projective flow algorithm proposed by Mann and Piccard in [20] to compute the homography parameters.

A homography $H = \{h_{ij}\}$, is a 3×3, 8 DOF projective transformation that models the relationship between the location of a feature at (x, y) in one frame, and the location (x', y') of the same feature in the next frame with eight parameters, such that,

$$x' = \frac{h_{11}x + h_{12}y + h_{13}}{h_{31}x + h_{32}y + 1}, \quad y' = \frac{h_{21}x + h_{22}y + h_{23}}{h_{31}x + h_{32}y + 1}. \tag{1}$$

The brightness constancy constraint results in a non-linear system of equations involving all pixels in the overlap range (region where the source and target images overlap). Using the method of [20], this system can be linearized for a least squares solution, such that, given two images, $I(x, y)$ and $I'(x, y)$, each pixel $i \in [1, N_p]$, then contributes an equation to the following system,

$$\begin{bmatrix} & x_i I_x(x_i, y_i) & \\ & y_i I_x(x_i, y_i) & \\ & I_x(x_i, y_i) & \\ & x_i I_y(x_i, y_i) & \\ \cdots & y_i I_y(x_i, y_i) & \cdots \\ & I_y(x_i, y_i) & \\ & x_i I_t(x_i, y_i) - x_i^2 I_x(x_i, y_i) - x_i y_i I_y(x_i, y_i) & \\ & y_i I_t(x_i, y_i) - x_i y_i I_x(x_i, y_i) - y_i^2 I_y(x_i, y_i) & \end{bmatrix}^{\top} \begin{bmatrix} h_{11} \\ h_{12} \\ h_{13} \\ h_{21} \\ h_{22} \\ h_{23} \\ h_{31} \\ h_{32} \end{bmatrix} = \begin{bmatrix} \vdots \\ x_i I_x(x_i, y_i) + y_i I_y(x_i, y_i) \\ -I_t(x_i, y_i) \\ \vdots \end{bmatrix}, \tag{2}$$

$$\mathbf{A}_{N_p \times 8} \mathbf{x}_{8 \times 1} = \mathbf{B}_{N_p \times 1}, \tag{3}$$

where $I_t(x_i, y_i) = I(x_i, y_i) - I'(x_i, y_i)$, $I_x(x_i, y_i) = \frac{\partial I(x_i, y_i)}{\partial x}$, and $I_y(x_i, y_i) = \frac{\partial I(x_i, y_i)}{\partial y}$, while h_{33} is 1. The least squares solution to this over-constrained system can be obtained with a singular value decomposition or pseudo-inverse. A coarse to fine estimation is achieved using three levels of Gaussian Pyramids. The spatial and temporal derivatives are also computed after smoothing using a Gaussian kernel of fixed variance. This process is fairly computation intensive as it involves solving a linear system of N_p equations where N_p is the number of pixels in each layer of the Gaussian pyramid. We used this technique as a baseline for comparison with our feature based registration algorithm in terms of speed and accuracy.

3.1.2 Feature-Based Method

As alternative to featureless gradient based methods, we study the performance of some feature-based alignment algorithms. We use two different algorithms to estimate homography with several types of feature detector algorithms. These two algorithms differ in the way they obtain correspondence between candidate feature-pairs of source and target images. Here is a detailed description of both the algorithms:

Flow based feature correspondence. In this algorithm, we extract invariant features from source image by applying one of the following methods:

- KLT [27] features. We obtain interest points in the image with significantly large eigenvalues by computing minimal eigenvalue for every source image pixel followed by non-maxima suppression in a local $d \times d$ neighborhood patch. Interest points with minimal value less than an experimentally determined threshold are eliminated prior to a final filtering based on spatial proximity of the features in order to extract only strong candidate interest points.

- SIFT [19] features. These features are extracted by computing the maxima and minima after applying difference of Gaussians at different scales. Feature points that lie along edges and points with low contrast are eliminated from the list of potential interest points. The dominant orientations are assigned to localized feature points. A 128-dimensional feature descriptor is obtained at each interest point extracted in this manner. We modify an open-source implementation of the SIFT algorithm[1] for extracting interest points.
- SURF [4] features. Speeded Up Robust Features are computed based on sums of responses obtained after applying a series of predefined 2-dimensional Haar wavelet responses on 5×5 image patches. The computation efficiency is enhanced up using integral images. A 128-dimensional vector is finally generated for each interest point.
- Random MSER [21] contour features. As the name suggests, we extract random points from contours returned after determining Maximally Stable Extremal Regions from an image. The MSERs are determined by first sorting image pixels according to their intensity, followed by a morphologically connected region merging algorithm. The area of each connected component is stored as a function of intensity. A larger connected component engulfs a smaller component until a maximally stable criterion is satisfied. Thus, MSERs are those parts of the image where local binarization is stable over a large range of thresholds.

Pixel locations corresponding to the features extracted using one of the above algorithms are stored in an $N \times 2$ matrix. These pixel locations are iteratively refined to find the interest point locations accurate to subpixel level. Using these sparse set of points from the source image, we compute respective optical flows in the target image. A pyramidal implementation [8] of Lucas Kanade's method is employed for this task which returns us corresponding points in the subsequent frame from the video sequence. A block diagram describing the important steps of this algorithm is shown in Fig. 2.

Descriptor similarity based feature correspondence. This algorithm works for those feature extraction methods that yield well defined descriptors for all detected interest points for e.g., SIFT and SURF, in a given source image. We first compute interest points in both source and destination images using either of the two methods. Thus we obtain two sets, which may not have equal number of interest points. Since each interest point is described in high-dimensional space, correspondences could be estimated using an approximate nearest neighbor search. We use a fast, freely available implementation [22] for this purpose. A block diagram is provided in Fig. 3 which explains this process. This technique is more robust as compared to the flow based mapping technique since it considers several attributes of the extracted feature points while generating the correspondence. However, it is computationally

[1] http://web.engr.oregonstate.edu/hess/downloads/sift/sift-latest.tar.gz

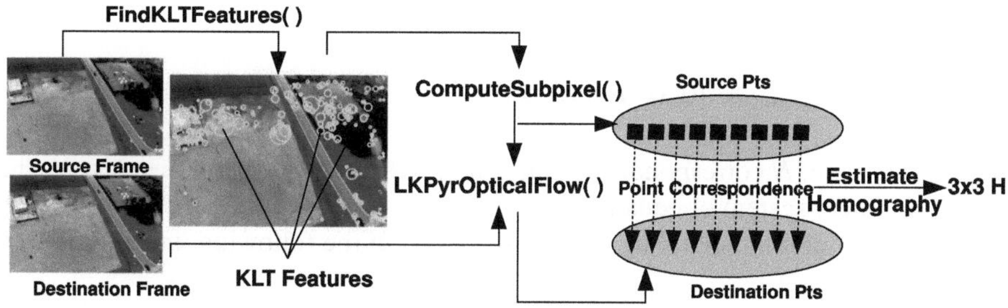

Fig. 2 Schematic diagram of the optical flow based correspondence mapping algorithm used for the motion compensation stage

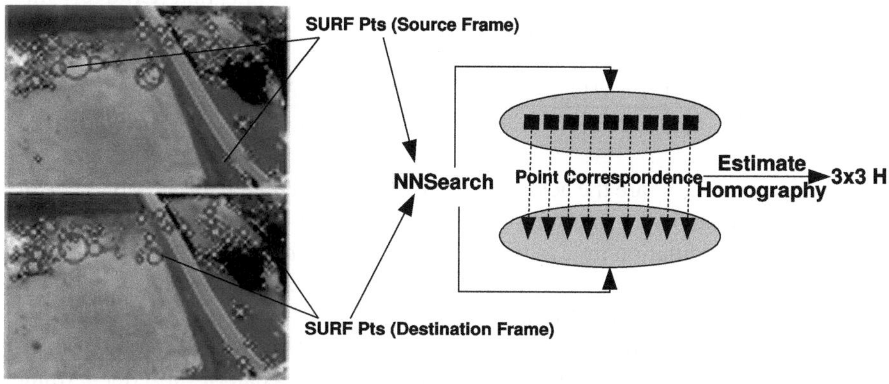

Fig. 3 Demonstration of the corresponding mapping algorithm based on descriptor similarity. This is used as an error correction mechanism in the cummulative homography computation step, within the motion compensation technique proposed here

more expensive than the former. We determine the accuracy of the registration algorithm by measuring the frame difference (FD) score. Formally, the FD score between a pair of consecutive intensity images I_t and I_{t+1} can be defined as:

$$FD = \frac{1}{N_p} \sum_{j=1}^{N_p} |I_t^j \times M(I_{t+1}^j) - W(I_{t+1}^j)|, \tag{4}$$

where $M(I_{t+1})$, $W(I_{t+1})$ are the outlier mask and the warped output of I_{t+1} with respect to I_t, respectively and N_p being the total number of pixels in a frame.

From the point correspondences established using either of the two methods discussed, we obtain respective pixel locations that are used to compute homography with the help of the following set of equations:

$$H = [h_{11}, h_{12}, h_{13}, h_{21}, h_{22}, h_{23}, h_{31}, h_{32}, h_{33}]^T, \tag{5}$$

$$a_x = [-x_i, -y_i, -1, 0, 0, 0, x_i'x_i, x_i'y_i, x_i']^T, \tag{6}$$

$$a_y = [0, 0, 0, -x_i, -y_i, -1, y_i'x_i, y_i'y_i, y_i']^T. \quad (7)$$

For a given set of N corresponding point pairs $\{(x_i, y_i), (x_i', y_i')\}$ for $1 \le i \le N$, the following linear system of equations hold good:

Given a set of corresponding points, we can form the following linear system of equations:

$$[a_{x_1}{}^T, a_{y_1}{}^T, a_{x_2}{}^T, a_{y_2}{}^T, \ldots, a_{x_N}{}^T, a_{y_N}{}^T]^T H = 0, \quad (8)$$

which is usually solved using random sampling technique [11] that iteratively minimizes the back-projection error, defined as:

$$\sum_i \left(x_i' - \frac{h_{11}x_i + h_{12}y_i + h_{13}}{h_{31}x_i + h_{32}y_i + h_{33}} \right)^2 + \left(y_i' - \frac{h_{21}x_i + h_{22}y_i + h_{23}}{h_{31}x_i + h_{32}y_i + h_{33}} \right)^2, \quad (9)$$

where x_i, y_i and x_i', y_i' are the actual and estimated 2D pixel locations and $h_{11} \ldots h_{33}$ are the nine elements of the homography matrix. It is interesting to note that the homography computation time in this case is significantly smaller than that observed in the feature-less method because the linear system formed here has significantly lesser number of equations than the former method.

The homography computed using the above methods reflects the transformation parameter from one frame to other and are only relative to a pair of subsequent frames. In order to have an understanding of the global camera motion, it is desired to obtain the transformation parameters of all subsequent frames with respect to the initial frame in the sequence. Therefore, we need to perform a cumulative multiplication of the homography matrices. Thus, the relative homography between image frame I_0 and I_n is

$$H_{0,n} = H_{0,1} \times H_{1,2} \times H_{2,3} \times \cdots \times H_{n-1,n}, \quad (10)$$

where, corresponding sets of points $\mathbf{x}_t$ and $\mathbf{x}_{t+1}$ in homogenous coordinates, for two frames I_t and I_{t+1}, can be related by

$$\mathbf{x}_{t+1} \approx H_{t,t+1}\mathbf{x}_t. \quad (11)$$

Now, for each of the cumulative homography matrix computed as above, we measure the curl and deformation metrics [28], using the following equations:

$$\text{curl} = |h_{12} - h_{21}|, \quad (12)$$

$$\text{deformation} = |h_{11} - h_{22}|. \quad (13)$$

These metrics are an approximate measure of the change in camera viewpoint in terms of camera orientation and translation. If either of these metrics are larger

```
Procedure CompensateMotion (V, k)
Input: Video Sequence (V)
Input: Number of frames to invoke error correction (k)
Output: Array of Cumulative Homographies H[]
count ← 0;
H[0] ← I;
src ← init frame;
src = EqualizeHistogram (src);
while not End of Sequence do
    dst ← next frame in sequence;
    dst = EqualizeHistogram (dst);
    if count is multiple of k then
        surfsrc = computeSURF (src);
        surfdst = computeSURF (dst);
        [srckpts, dstkpts] ← findNearestNeighbor (surfsrc, surfdst);
    else
        kpts = findKLTFeatures (src);
        srcpts = computeSubpixel (kpts);
        dstpts = LKPyrOpticalFLow (src, dst, srcpts);
    h ← RANSACFitHomography (srckpts, dstkpts);
    H[count + 1] = h * H[count];
    curl ← computeCurl(H[count+1]);
    def ← computeDef(H[count+1]);
    if curl > CURL_THRES or def > DEF_THRES then
        H[count + 1] ← I;
    result ← warpProjective (dst, H[count]);
    src ← dst;
    count ← count +1;
```

Algorithm 1 Pseudo-code describing the motion compensation algorithm used by COCOA-LIGHT on FLIR imagery with KLT features for establishing flow based correspondence and SURF used to regulate cumulative homography drift

than an empirical threshold, the consecutive frames indicate a significant change in view-point, and therefore a higher likelihood of erroneous alignment. Under these circumstances we reset the relative homography matrix to identity and frames from here on are treated as a new sub-sequence.

However, the cumulative homography computation as discussed is not robust to errors. Slight noise in the estimation of homography in one pair of frames can be easily propagated through the cumulative homography matrix resulting in errors that could affect the overall accuracy of the motion compensation, thereby causing errors in the object detection stage. In order to alleviate the effect of such erroneous calculations, we introduce a small error correction measure after every K frames, where the cumulative homography is replaced with homography estimated directly from descriptor mapping. This enhances the overall accuracy with the cost of a slight computation overhead. The results of applying motion compensation on an example three vehicle sequence are shown in Fig. 4. Each image in the figure is generated by allocating the first two channels of an RGB image matrix with reference frame

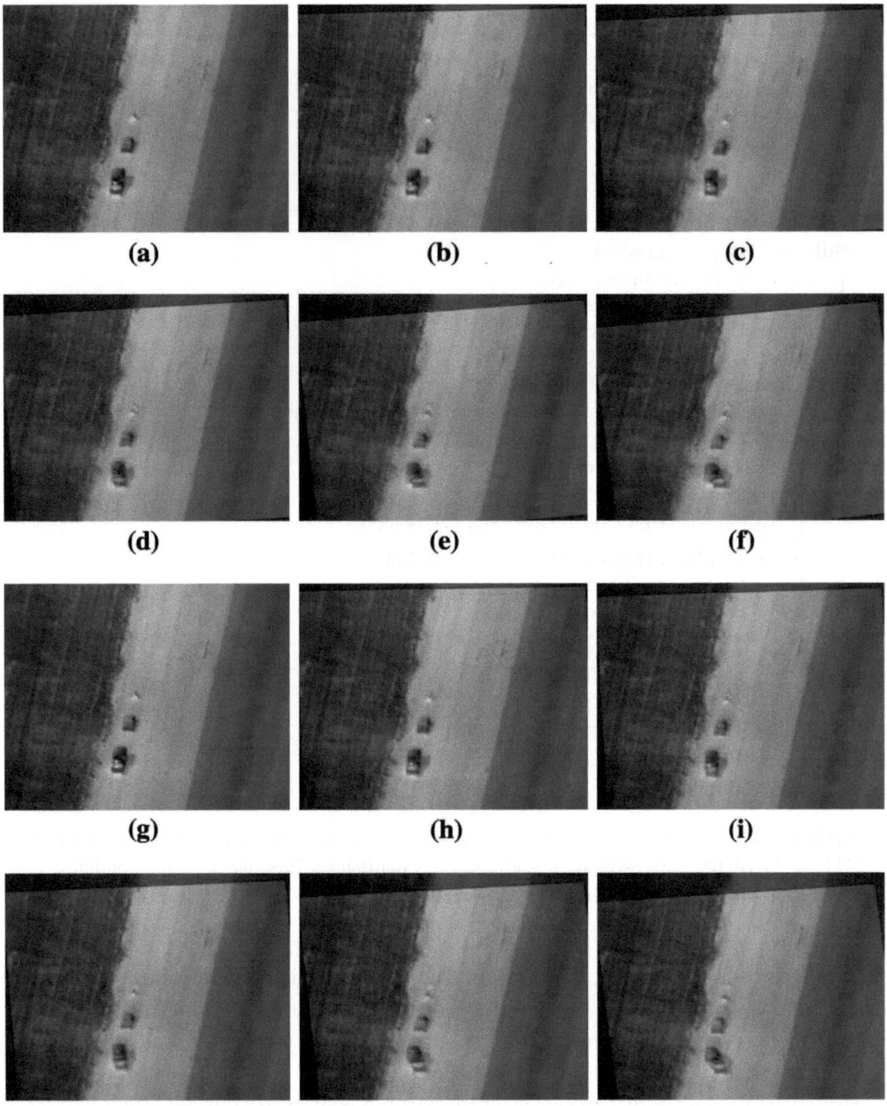

Fig. 4 Comparing alignment using cummulative homography computed using gradient based and KLT-feature based methods: images labeled **a–f** are aligned using the gradient feature based registration algorithm while images from **g–l** are aligned using the KLT-feature based algorithm. The real-valued number in parentheses corresponding to each image is its normalized frame difference scores obtained by subtracting aligned destination frame from the initial frame in the sequence. A smaller score indicates a better accuracy in alignment. **a** Frame 0 (0.0000), **b** Frame 0–10 (1.0110), **c** Frame 0–20 (1.1445), **d** Frame 0–30 (1.6321), **e** Frame 0–40 (1.9821), **f** Frame 0–50 (2.3324), **g** Frame 0 (0.0000), **h** Frame 0–10 (0.9121), **i** Frame 0–20 (1.1342), **j** Frame 0–30 (1.5662), **k** Frame 0–40 (1.8995), **l** Frame 0–50 (2.3432)

(a) (b)

Fig. 5 Global mosaics generated after image alignment are shown for **a** the three vehicle sequence, and **b** the distant view sequence

and its subsequent aligned counterpart in grayscale, respectively. Regions that do not align properly are visible as green patches. Hence, in a set of correctly motion compensated frames, the green patches correspond to the moving objects as evident in Fig. 4. Global mosaics corresponding to the two different sequences discussed in this paper are shown in Fig. 5a and b. The complete motion compensation algorithm is listed in Algorithm 1.

With this knowledge, we proceed to our next section that discussed the methods we have employed to detect moving objects from a set of ego-motion compensated frames.

3.2 Object Detection

Given a sequence of frames, the goal of object detection is to obtain blobs for foreground objects. Background subtraction is a popular approach for static cameras where the background at each pixel can be modeled using mean, median, Gaussian, or a mixture of Gaussians. In aerial videos, background modeling is hindered due to camera motion. Although the aligned frames seem visually similar to a sequence of frames from a static camera, there are marked differences at the pixel level where errors in alignment cause small drifts in the pixel values. Such drifts are more pronounced near sharp edges. Furthermore, these drifts can be in different directions in different parts of the scene for each frame.

The most significant amongst the issues that pose challenge to background modeling in aerial videos are the errors due to parallax. Since we use features-based alignment, there are many features which come from out-of-plane objects such as buildings and trees. These features affect the computation of homography which is computed using all feature correspondences between a pair of consecutive frames.

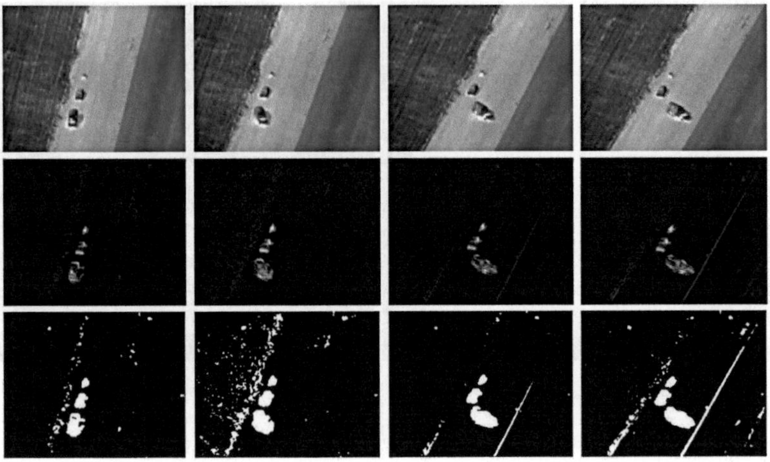

Fig. 6 The *first row* shows the original frames (10, 50, 100 and 150) from Sequence 1 while the *second* and *third rows* show accumulative frame difference (AFD) and AFD after thresholding respectively

This is the inherent source of error whose effect is visible near high gradients in a frame. Since all the homographies are computed between consecutive frames, the error in alignment accumulates with time. Even if we choose a small yet reasonable number of frames for constructing the background, the drift in the scene due to accumulated errors hampers the computation of background. (See discussion for Fig. 18).

Another reason is the limitation on the number of frames available for modeling the background. A region has to be visible for a reasonable number of frames to be learned as background. In the case of a moving camera, the field-of-view changes at every frame which puts restraints on the time available for learning. If the learning time is too short, some pixels from foreground are modeled as background. A constant change in field-of-view is also the reason that it is not possible to choose a single reference frame when performing alignment doing which can allow us to get rid of accumulated errors. After a few frames, the field-of-view of the new frame might not overlap with that of the reference frame and will thus disallow the computation of homography.

In addition to the two issues mentioned above, background modeling is computationally expensive for registered images which are usually greater in size than the original frames, and is thus prohibitive for longer sequences. In order to make foreground detection close to real time and cater for the non-availability of color information in FLIR imagery, we use a more feasible alternative of accumulative frame differencing (AFD), which takes as input only a neighborhood of n frames for detection at each time step (Fig. 6).

For each frame, the algorithm is initialized using a constant number of frames called temporal window of size of $2n + 1$ with n frames on both sides of the

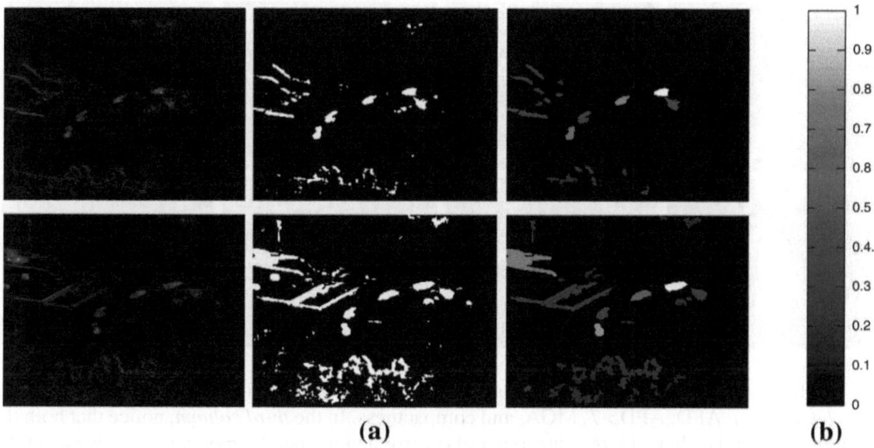

Fig. 7 **a** The *first column* shows AFD for two frames from Distant view Sequence whereas *second column* shows AFD after thresholding. As can be seen from the *third column*, mean gray area (normalized between 0 and 1) of blobs corresponding to moving objects is high which can be used to separate moving objects from noisy blobs. **b** shows the *gray-map* used for all the figures in this section

current frame. This means the detection procedure will have a lag of n frames. The accumulative frame difference for ith frame (I_i) for temporal window from $-n$ to n is given by

$$\mathrm{AFD}(I_i, n) = \sum_{k=i-n}^{i+n} |I_i - W(I_i, I_k)|, \tag{14}$$

where $W(I_i, I_k)$ is a function to warp kth frame to the ith frame.

We experimented with different size of temporal window with the conclusion that $n = 10$ is empirically the most suitable value. If n is close to 2 , the blobs are small, incomplete and missing. If we go beyond 10, the blobs start to merge and sharp edges of the background begin to appear as false positives.

The grayscale image obtained after accumulative frame differencing is normalized between 0 and 1 followed by thresholding (with discardThreshold T). Blobs are obtained using connected-component labeling. Since pixels belonging to moving objects have higher values in accumulative frame difference than noise (see Fig. 7), mean gray area of such blobs is correspondingly high. Moreover, it can be observed that blobs corresponding to moving objects are compact and regular in shape when compared against irregular shaped blobs due to noise (see Fig. 8). However, an exception to this are the noisy blobs that come from regions of high gradients some of which might not be irregular in shape. Instead, they have a prominent characteristic of being elongated with higher eccentricity. Figure 9 explains the use of eccentricity as a measure to cater for such blobs.

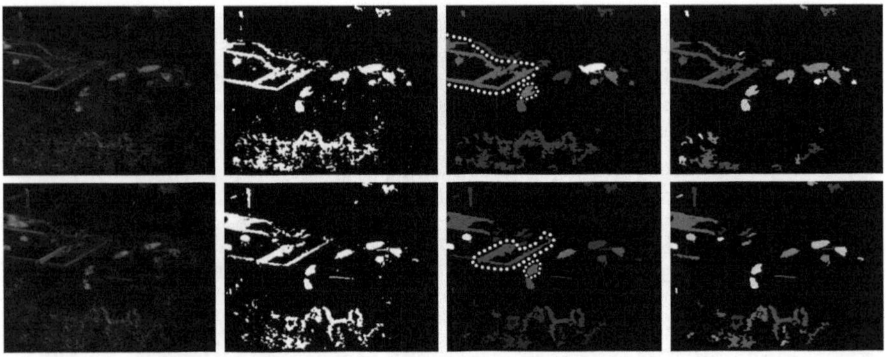

Fig. 8 This figure illustrates the advantage of using compactness for removing false positives. From *left* to *right*: AFD, AFD > T, MGA, and compactness. In the *third column*, notice that both the highlighted irregular shaped blobs due to parallax error and the nearby moving object have similar MGA. However, blobs due to moving objects are more compact (*fourth column*) and will therefore get higher weight

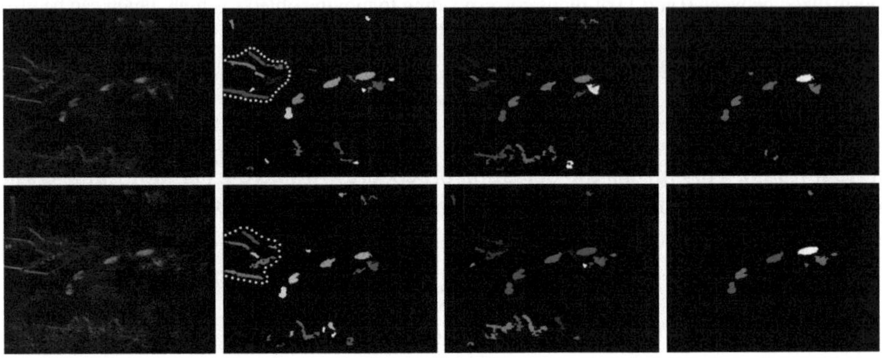

Fig. 9 From *left* to *right*: AFD, compactness, eccentricity and weights of final blobs. The highlighted elongated blobs due to noise do not get suppressed using compactness in the *second column* but do get lower eccentricity weight as shown in the *third column*

We will now give definitions for the three measures. If $b_t^i \in \mathbf{B}_t$ denotes the ith blob at frame t, then its mean gray area, compactness and eccentricity are computed using the following formula:

$$\text{Mean Gray Area}^i = \frac{\sum_{\forall p(x,y) \in b_t^i} \text{AFD}(x, y)}{|b_t^i|}, \tag{15}$$

$$\text{Compactness}^i = \frac{|P(b_t^i)|}{2\pi\sqrt{|b_t^i|/\pi}}, \tag{16}$$

$$\text{Eccentricity}^i = \sqrt{\frac{2C_{xy}}{u_{xx} + u_{yy} + C_{xy}}}, \tag{17}$$

where P gives perimeter of the blob. u_{xx}, u_{yy} and C_{xy} are given by

$$u_{xx} = \frac{\sum_{\forall p(x,y)\in b_t^i} (x - \bar{x})^2}{|b_t^i|} + \frac{1}{12}, \quad u_{yy} = \frac{\sum_{\forall p(x,y)\in b_t^i} (y - \bar{y})^2}{|b_t^i|} + \frac{1}{12} \tag{18}$$

$$C_{xy} = \sqrt{(u_{xx} - u_{yy})^2 + 4u_{xy}} \quad \text{where } u_{xy} = \frac{\sum_{\forall p(x,y)\in b_t^i} (x - \bar{x})(y - \bar{y})}{|b_t^i|} \tag{19}$$

where 1/12 is the normalized second central moment of a pixel with unit length.

The following equation describes the scheme to combine weights from mean gray area, compactness and eccentricity:

$$W^i = \alpha_1 \times \text{MGA}^i + \alpha_2 \times (2 - \text{Compactness}^i) + \alpha_3 \times (1 - \text{Eccentricity}^i), \tag{20}$$

where α_1, α_2, and α_3 are empirically determined constants with relatively higher weight given to MGA. The blobs are sorted according to their weights W^i and normalized between 0 and 1 and only $\min(\text{maxObjects}, \ |b_t^i| \ |W^i > T)$ are returned where maxObjects is a hard limit on the maximum number of output objects. The reason AFD and W^i are normalized by their respective maximum values is to keep T constant across different sequences. The empirical value of discardThreshold T is .005 or .5% of maximum value. If T is too low for some frame, it can cause blobs from moving objects to merge with those from the noise (see Fig. 10). Since pixels from high motion objects will have higher values in AFD, all such pixels should be output as foreground. If the detection procedure discards pixels that should have been included in output, T is progressively increased till all high motion objects are included in the output.

Though the proposed approach gives reasonable results across a wide variety of sequences without changing any weights and the threshold T, information regarding minimum and maximum blob size can be incorporated in Eq. 20 to fine tune the results for a particular configuration of camera altitude and scene clutter. Figure 19 provides intermediate for the detection in three frames from Distant View Sequence.

We evaluate the performance of our detection algorithm, using Multiple Object Detection Precision (MODP) [6] scores in addition to the standard Probability of Detection (PD) and False Alarm Rate (FAR) metrics from Automatic Target Recognition literature [23]. The MODP is calculated on a per frame basis and is given as:

$$\text{MODP}_t = \frac{1}{N_t} \sum_{i=1}^{N_t} \frac{|G_t^i \cap B_t^i|}{|G_t^i \cup B_t^i|}, \tag{21}$$

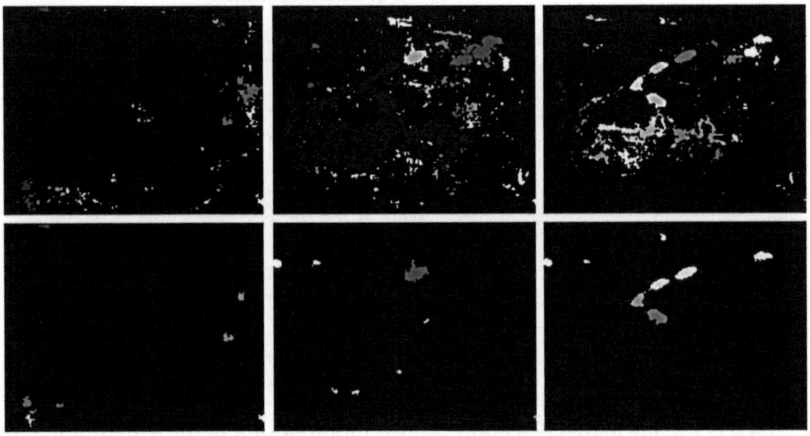

Fig. 10 Progressive thresholding: *top row* shows the connected component labels obtained with discardThreshold = .5%, .8% and 1%. The invisible *top-left* region corresponds to blobs with smaller labels (close to zero). *Bottom row* depicts the corresponding detections. discardThreshold is progressively increased from .5% to 1% till the objects and noise form separate blobs

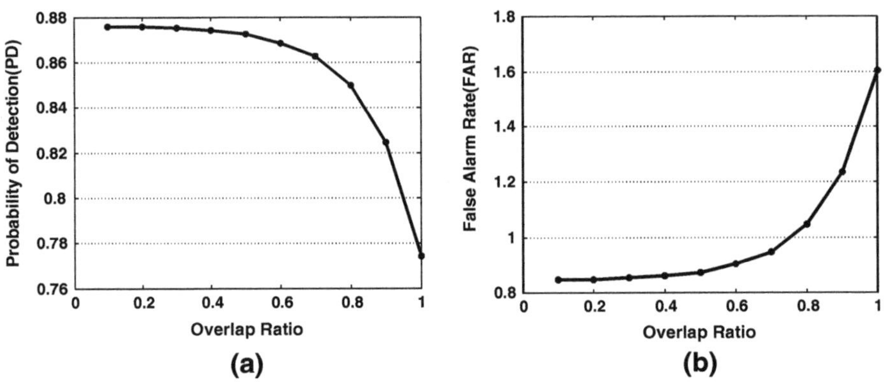

Fig. 11 Detection evaluation obtained on Distant View Sequence with varying overlap ratio from 0.1 to 1. **a** Probability of detection scores, **b** false alarm rate

where B_t and G_t are the respective set of corresponding objects output by the detection stage and that present in Ground Truth, at frame t, N_t being the cardinality of the correspondence. The fractional term in Eq. 21 is also known as the spatial overlap ratio between a corresponding pair of bounding boxes of ground-truthed and detected objects. Figure 11 reports the PD and FAR scores for Distant View Sequence, obtained by varying the bounding box overlap ratio.

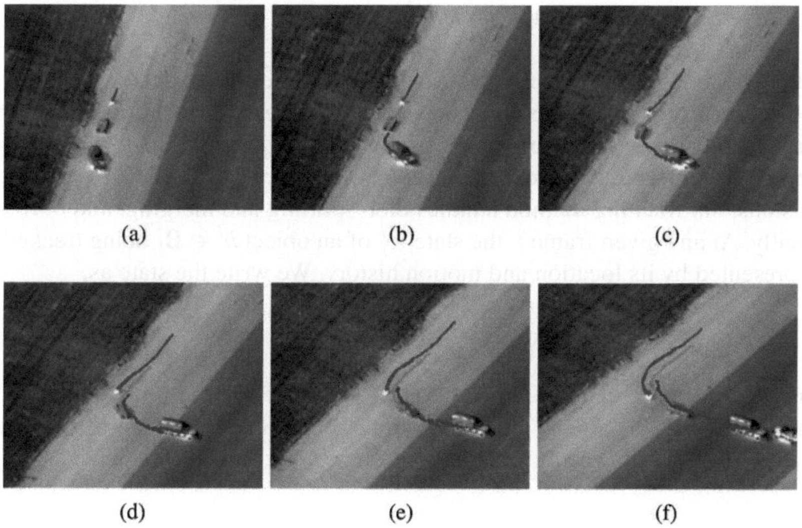

Fig. 12 Tracking results for three vehicle sequence. Tracks of multiple objects are overlaid on every 50th frame. All three visible objects are tracked correctly for the duration of the sequence. A fourth object just entering the camera's field of view is visible in frame 300. **a** Frame 50, **b** Frame 100, **c** Frame 150, **d** Frame 200, **e** Frame 250, **f** Frame 300

3.3 Tracking

The process of object detection provides a set of unique labels assigned to mutually exclusive groups of pixels for each image, where each label ideally corresponds to a single moving object. Given that the set of observed objects is denoted by $\mathbf{B}_t = \{b^i\}$, where $1 \leq i \leq O_t$, and O_t is the number of objects detected in frame t, the problem of tracking is defined as computation of a set of correspondences that establishes a 1–1 relationship between $b^i \in \mathbf{B}_t$ for all i, with an object $b^j \in \mathbf{B}_{t+1}$. In addition to problems like occlusions, non-linear motion dynamics, and clutter, that are traditionally encountered in object tracking in static, surveillance cameras, tracking in aerial FLIR imagery is made much harder because of low image resolution and contrast, small object sizes, and artifacts introduced in images during the platform motion compensation phase. Even small errors in image stabilization can result in a significant number of spurious object detections, especially in regions with high intensity gradients, further complicating the computation of the optimal object correspondence across frames (Fig. 12).

3.3.1 Kinematic Constraint

Our tracking algorithm employs a constant velocity motion model, and various cues for object correspondence, including appearance, shape and size. Furthermore, due to severe splitting and merging of objects owing to potential errors in detection, as well as apparent entry and exit events due to object to object, and object to background occlusions, our tracking method handles blob splitting and merging, and occlusions explicitly. At any given frame *t*, the state X_t^i of an object $b^i \in \mathbf{B}_t$ being tracked, can be represented by its location and motion history. We write the state as,

$$X_t^i = [x_t^i, y_t^i, \rho_t^i, \theta_t^i], \tag{22}$$

where (x^i, y^i) represents the 2d location of the object on the image plane at time (frame) *t*, and (ρ^i, θ^i) are the magnitude and orientation of the mean velocity vector of the object. The state vector for object *i* at frame $t+1$, X_{t+1}^i is predicted as follows:

$$\begin{bmatrix} x_{t+1} \\ y_{t+1} \end{bmatrix} = \begin{bmatrix} x_t \\ y_t \end{bmatrix} + \begin{bmatrix} \rho_t \cos\theta_t \\ \rho_t \sin\theta_t \end{bmatrix} + \begin{bmatrix} \gamma_x \\ \gamma_y \end{bmatrix}, \tag{23}$$

where (γ_x, γ_y) depict Gaussian process noise with zero mean and standard deviations σ_x and σ_y in *x* and *y* directions, which are derived from the variation in (ρ, θ) over time (the correlation is assumed to be zero). Assuming the magnitude and orientation of the velocity vector between an object's location at time frame *t* and $t-1$ to be $\hat{\rho}_t$ and $\hat{\theta}_t$ respectively, the velocity history in the state vector is updated by computing the weighted means of object's velocity magnitude and orientation in the current and previous frames, i.e., ρ_t and $\hat{\rho}_t$. The orientation of the object's velocity is similarly updated, by phase change invariant addition, subtraction and mean functions.

The motion model based probability of observing a particular object with state X_t^i in frame *t*, as object $b^j \in \mathbf{B}_{t+1}$ with centroid (x_{t+1}^j, y_{t+1}^j) in frame $t+1$ can then be written as

$$P_m(b_{t+1}^j | X_t^i) = \frac{1}{2\pi\sigma_x\sigma_y} \exp\left\{ -\frac{1}{2}\left[\frac{(x_{t+1}^i - x_{t+1}^j)^2}{\sigma_x^2} + \frac{(y_{t+1}^i - y_{t+1}^j)^2}{\sigma_y^2} \right] \right\}. \tag{24}$$

Notice that we can compute (x_{t+1}^i, y_{t+1}^i) from the constant velocity motion model as described before.

3.3.2 Observation Likelihood

In addition to the motion model described above, the key constituent of correspondence likelihood between two observation in consecutive frames is the observation model. Various measurements can be made from the scene to be employed for use in observation model, which combined with the kinematics based prediction defines the cost of association between two object detections. As described earlier, we used appearance, shape and size of objects as measurements. These observations for an object denoted by b^i are denoted by δ_c^i, δ_g^i, δ_s^i, and δ_a^i, for intensity histogram, mean gray area (from frame difference), shape of blob, and pixel area of the blob respectively. The probability of association between two blobs using these characteristics can then be computed as follows. $P_c(b_{t+1}^j|X_t^i)$ denotes the histogram intersection between histograms of pixels in object's bounding box in the previous frame, and the detection under consideration, i.e., b_{t+1}^j.

The probability $P_g(b_{t+1}^j|X_t^i)$ can simply be computed using the difference in the mean gray values of each blob after frame differencing, normalized by maximum difference possible. The shape based likelihood is computed by aligning the centroids of blobs b_t^i and b_{t+1}^j, and computing the ratio of blob intersection and blob union cardinalities, and is represented by $P_s(b_{t+1}^j|X_t^i)$. Finally the pixel areas for the blobs can be compared directly using the variance in an object's area over time which is denoted by σ_a^i. The probability of size similarity is then written as, $P_a(b_{t+1}^j|X_t^i) = \mathcal{N}(\delta_a^j|\delta_a^i, \sigma_a^i)$, where $\mathcal{N}$ represents the Normal distribution.

Assuming the mutual independence of motion, appearance, shape and size, we can write the probability of a specific next object state (the blob detection b_{t+1}^j), given all the observations, to be,

$$\begin{aligned} P(b_{t+1}^j|X_t^i, \delta_c^i, \delta_g^i, \delta_s^i, \delta_a^i)X = {} & P_m(b_{t+1}^j|X_t^i)P_c(b_{t+1}^j|X_t^i)P_g(b_{t+1}^j|X_t^i) \\ & \times P_s(b_{t+1}^j|X_t^i)P_a(b_{t+1}^j|X_t^i), \end{aligned} \tag{25}$$

which gives the aggregate likelihood of correspondence between the blob $b_t^i \in \mathbf{B}_t$ in frame t represented by state X_t^i, and the blob $b_{t+1}^j \in \mathbf{B}_{t+1}$ in frame $t+1$.

3.3.3 Occlusion Handling

Tracking in traditional surveillance scenarios and especially in aerial FLIR imagery suffers from the problems of severe object to object and object to background occlusions. Furthermore, the low resolution and low contrast of these videos often induce high similarity between objects of interest and their background, thus resulting in mis-detections. Consequently, a simple tracker is likely to initialize a new track for an object undergoing occlusion every time it reappears. To overcome this problem, our tracking algorithm continues the track of occluded object by adding hypothetical points to the track using its motion history. In actuality, the track of every object in the current frame, that does not find a suitable correspondence in the next frame, within an ellipse defined by five times the standard deviations σ_x and σ_y, is propagated using this method. In particular, it is assumed that the occluded object will maintain persistence of appearance, and thus have the same intensity histogram, size, and shape. Obviously, according to the aggregate correspondence likelihood, such a hypothetical object will have nearly a 100% chance of association. It should be noted however that an implicit penalty is associated with such occlusion reasoning that arises from the probability term $P_g(\cdot)$, which in fact can be computed regardless of detection. In other words, the mean gray area of the hypothetical blob (deduced using motion history) is computed for the frame in question, which reduces the overall likelihood of association as compared to an actual detected blob which would have a relatively low likelihood otherwise. This aggregate probability is denoted by $P_o(b^{\hat{k}}_{t+1}|X^k_t)$, where $b^{\hat{i}}_{t+1}$ is the hypothetical blob in frame $t+1$, resulting from motion history based propagation of the blob b^i_t described by the state vector X^i_t. The track of an object that has exited the camera view can be discontinued by either explicitly testing for boundary conditions, or by stopping track propagation after a fixed number of frames.

3.3.4 Data Association

Given blobs in consecutive frames t and $t+1$ as $\mathbf{B}_t$ and $\mathbf{B}_{t+1}$, their state and measurement vectors, probability of association between every possible pair of blobs is computed. The goal of the tracking module then is to establish 1–1 correspondence between the elements of the sets $\mathbf{B}_t$ and $\mathbf{B}_{t+1}$. Numerous data association techniques have been proposed in the computer vision literature, including methods for single, few, or a large number of moving targets. Many of these methods (e.g., bipartite graph matching) explicitly enforce the 1–1 correspondence constraint, which may not be ideal in the FLIR sequences scenario, since a non-negligible number of false positive and false negative detections can be expected.

We, therefore, employ an object centric local association approach, rather than a global association likelihood maximization. This technique amounts to finding the nearest measurement for every existing track, where 'nearest' is defined in the observation and motion likelihood spaces (not the image space). This approach is also known as the greedy nearest neighbor (GNN) data association [7]. Formally, for the trajectory i, containing the measurement $b^i_t \in \mathbf{B}_t$, described by the current state X^i_t, the next associated measurement can be computed as

$$b^i_{t+1} = \underset{j\in[1,O_{t+1}]}{\operatorname{argmax}} P(b^j_{t+1}|X^i_t, \delta^i_c, \delta^i_g, \delta^i_s, \delta^i_a). \tag{26}$$

The objects in the set $\mathbf{B}_{t+1}$, that are not associated with any existing track can be initialized as new trajectories, while existing tracks not able to find a suitable correspondence are associated with a hypothetical measurement as described earlier. If a track cannot find real measurements after addition of a predetermined number of hypothetical blobs, the track is discontinued.

The performance of the tracking algorithm discussed here is evaluated using a metric similar to the one shown in Eq. 21. Multiple Object Tracking Precision (MOTP) is given by

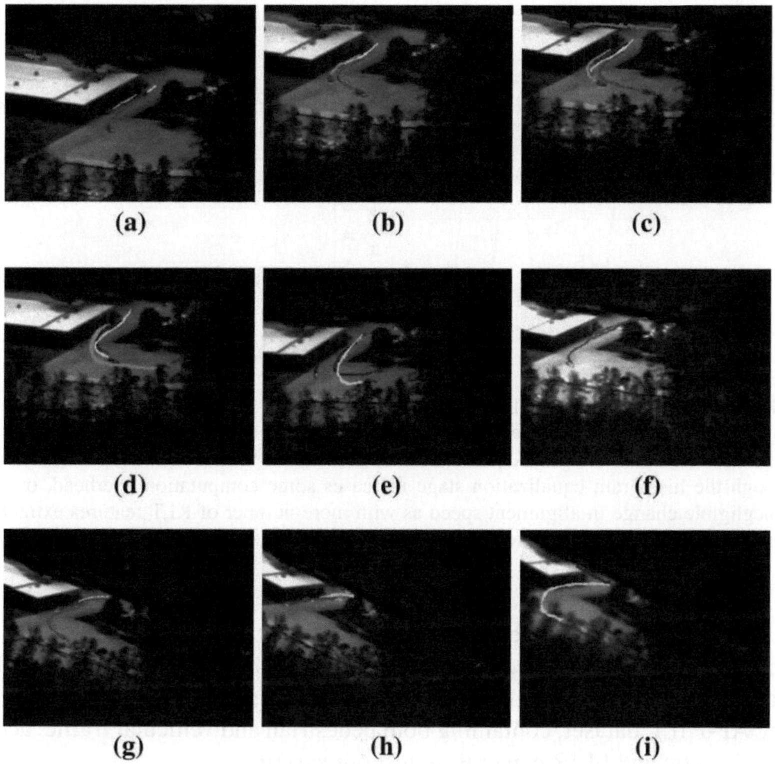

Fig. 13 Tracking of vehicles in distant field of view. Tracks of multiple objects are overlaid on frames of the sequence at regular intervals. The same *gray-scale* indicates consistent labeling of the object. Most of the objects are tracked throughout their observation in the camera's field of view. Notice the low resolution and contrast. **a** Frame 56, **b** Frame 139, **c** Frame 223, **d** Frame 272, **e** Frame 356, **f** Frame 422, **g** 500, **h** 561, **i** 662

$$\mathrm{MOTP}_t = \frac{\sum_{i=1}^{N_t} \sum_{t=1}^{N_f} \left[\frac{|G_t^i \cap B_t^i|}{|G_t^i \cup B_t^i|} \right]}{\sum_{j=1}^{N_t} N_t^j} \tag{27}$$

where N_t refers to the mapped objects over an entire trajectory as opposed to a single frame. The MOTP scores for a subset of 12 sequences are shown in Fig. 13, in the following section.

4 Discussion

In this section we provide an in-depth analysis of the various algorithms that are used in cocoalight in terms of their individual performance followed by an overall execution summary of the system. All the following experiments are conducted on a

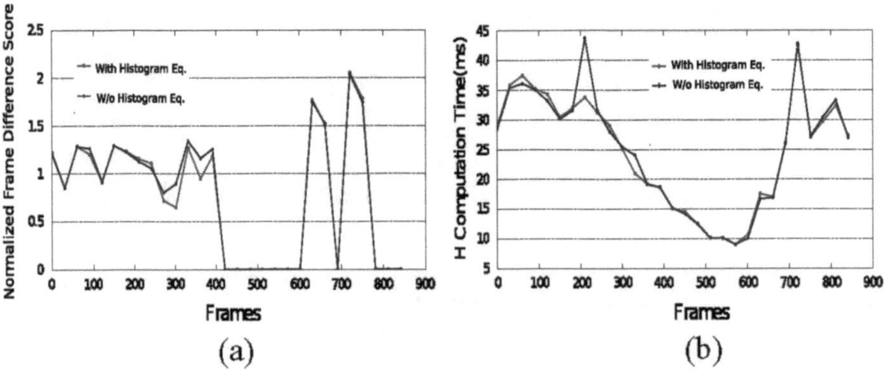

Fig. 14 Effect of histogram equalization on the accuracy of alignment and computation time. **a** Accuracy achieved in alignment after histogram equalization three vehicle sequence. The results shown here indicate that histogram equalization is beneficial for feature extraction in FLIR imagery. **b** Although the histogram equalization stage increases some computation overhead, overall we notice negligible change in alignment speed as with more number of KLT features extracted, the homography estimation routine takes fewer RANSAC iterations to generate optimal solution

desktop computing environment with a 1.6 GHz Intel x86 dual core CPU and 2 GB physical memory. The two sequences containing vehicular traffic, shown earlier in this paper are acquired from the VIVID 3 dataset. In addition, we use a more challenging AP-HILL dataset, containing both pedestrian and vehicular traffic, acquired by Electro-optic and FLIR cameras, to test our system.

A quantitative improvement in alignment accuracy and computational performance due to contrast enhancement is shown in Fig. 14. It can be noted that the total frame difference per frame is reduced after alignment using histogram equalization, due to an increased number of relevant feature points in regions of previously low contrast. On the other hand, this process is not a computational burden on the system, and in some cases can even improve the transformation computation time. In Fig. 15, we analyze the drift or error in estimation that is introduced in the cumulative homography computation stage. For the sake of simplicity, we only show the results corresponding to the parameters that only determine translation across frames in a sequence. We observe that curves corresponding to either parameters, have similar slopes which indicates that the proposed algorithm 1 achieves results closer to the gradient based method. It is worthwhile to note that our algorithm is more robust to change in background than the gradient based method as it has lesser number of homography reset points (where the curves touch the x-axis).

Figure 16 summarizes the impact of increasing the number of KLT features in the motion compensation stage. As the number of features are increased, we observe a drop in the computation speed in Fig. 16b. The accuracy in alignment, which is measured in terms of normalized frame difference scores, however shows marginal improvement beyond 512 features. In a slightly different setting, we evaluate different types of feature extraction strategies against the gradient based method. In

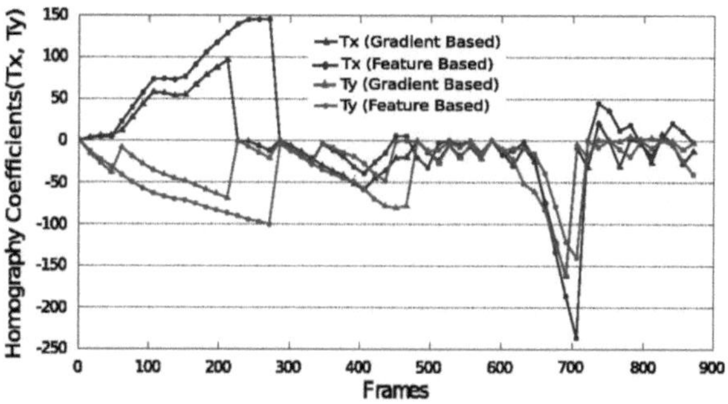

Fig. 15 Comparing homography parameters estimated using KLT feature based method against the gradient-based method. Parameters corresponding to the translation along x and y axes, represented by *curves of different gray-scale values*. It is interesting to observe the frame locations along x-axis where the parameter curves touch the x-axis. These locations indicate the positions when the homography is reset to identity because of large frame motion

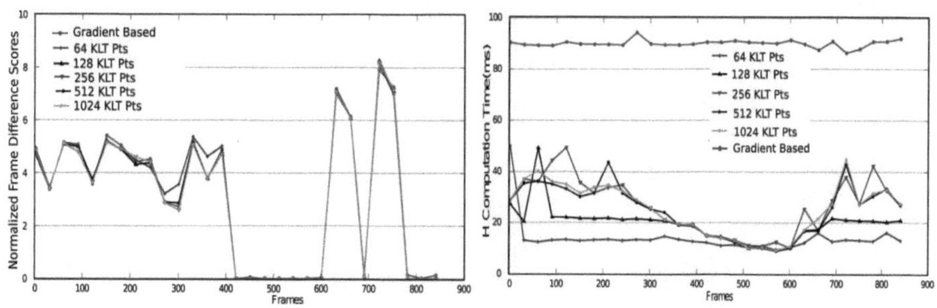

(a) (b)

Fig. 16 Effect of increasing KLT features on alignment: **a** accuracy achieved in alignment on three vehicle sequence with different number of KLT features. As number of features are increased, the alignment accuracy reaches close to that achieved using gradient based method. **b** Computation time of homography is maximum with gradient based method and reduces significantly with decrease in number of KLT features

Fig. 17a, we notice that both KLT and SIFT feature based methods achieve accuracies comparable to the gradient based scheme with the KLT feature based method being twice as computationally efficient as the gradient and SIFT feature based methods.

The alignment algorithm used by Cocoalight makes a planarity assumption on the input scene sequences. This implies that pixels from ground plane, that contribute to the linear system of equations for computing homography, should outnumber those from outside the ground plane. If this criterion is not satisfied, homography between two frames cannot be computed accurately. This is usually observed in typical urban scenarios that consist of tall buildings imaged by low flying UAVs. We demonstrate this issue in Fig. 18. The alignment error is largely visible as we proceed towards the end of the sequence in Fig. 18c.

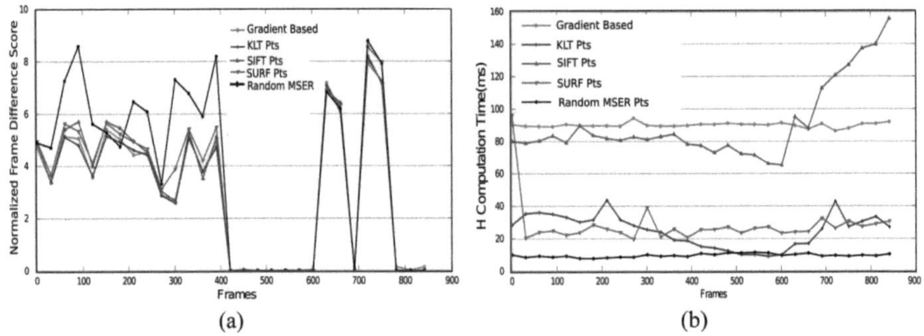

Fig. 17 Effect of different types of features on alignment: **a** accuracy achieved with different feature extraction algorithms (KLT, SIFT, SURF, MSER) in comparison to the gradient based method, and **b** their respective homography computation time

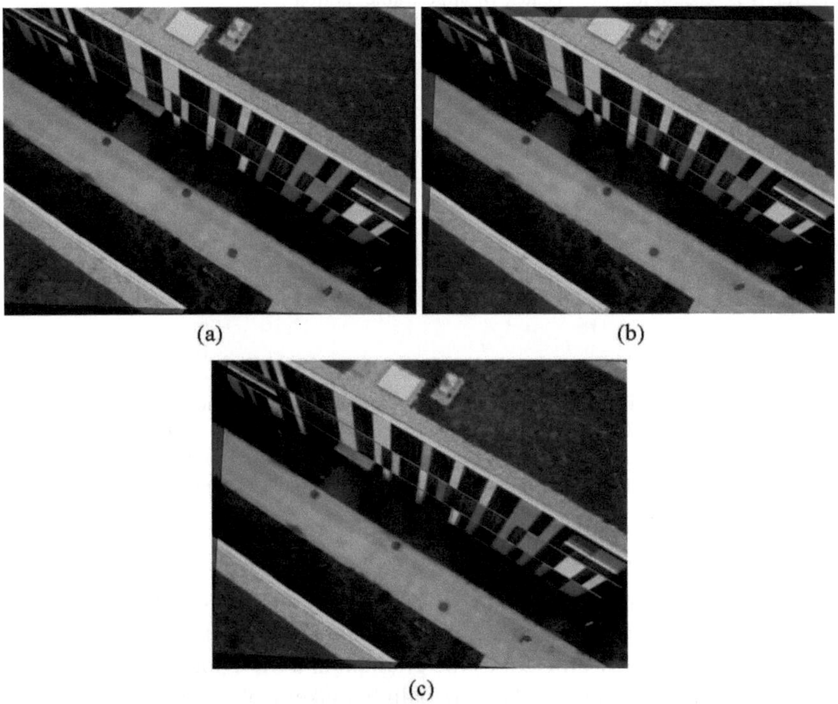

Fig. 18 Erroneous alignment due to pixels outside the ground plane contributing in homography estimation. **c** *Green* visible patches near the circular drainage holes did not align properly. **a** Frame 0/20, **b** Frame 0/40, **c** Frame 0/60

In Table 1, we report the performance of our detection and tracking setup against different evaluation metrics, namely PD, MODP, MOTP and FAR for a subset of 12 sequences from our datasets. These sequences are characterized by

Table 1 Quantitative evaluation of runtime for individual modules, namely Motion compensation (alignment), ROI detection and Tracking for 12 FLIR aerial sequences from the AP-HILL dataset containing moving vehicles and human beings

Sequence	Frames	Alignment	Detection	Tracking	FDA	PD	FAR	MOTP	MOTA
Seq. 01	742	23.3	8.3	36.1	4.89	0.81	0.12	0.67	0.74
Seq. 02	994	21.6	7.9	39.6	6.77	0.89	0.08	0.69	0.71
Seq. 03	1138	24.0	6.1	38.1	10.89	0.88	0.09	0.65	0.76
Seq. 04	1165	22.2	6.5	40.6	11.32	0.78	0.05	0.69	0.81
Seq. 05	1240	24.3	9.4	40.2	4.22	0.83	0.13	0.75	0.82
Seq. 06	1437	25.1	6.2	41.0	7.95	0.91	0.06	0.63	0.69
Seq. 07	1522	21.4	8.3	36.7	6.83	0.87	0.04	0.61	0.78
Seq. 08	1598	25.6	7.9	38.2	5.39	0.76	0.06	0.64	0.75
Seq. 09	1671	24.8	6.1	36.1	7.94	0.73	0.11	0.61	0.74
Seq. 10	1884	22.8	6.1	42.1	8.83	0.75	0.09	0.59	0.78
Seq. 11	1892	23.6	6.7	39.4	12.56	0.82	0.12	0.66	0.69
Seq. 12	1902	21.7	8.4	41.5	10.21	0.89	0.06	0.72	0.73

Each video sequence has a spatial resolution of 320×240 and are arranged in ascending order of the number of frames contained in them for better readability. The Frame Difference Score averaged over the total number of frames in a given sequence serves as the performance metric for the alignment module. Probability of Detection (PD) and False Alarm Rate (FAR) measures provide vital insights on the performance of the detection module. Finally, Multiple Object Tracking Precision (MOTP) and Multiple Object Tracking Accuracy (MOTA) scores are presented for each of these sequences to measure the performance of the Tracking module

the following: (a) small and large camera motion, (b) near and distant field of views, (c) varying object sizes (person, motorbike, cars, pick-ups, trucks and tanks), (d) background clutter.

Some qualitative tracking results for near and far field sequences are shown in Figs. 12 and 13 respectively. Object tracks are represented as trajectories, which are lines connecting the centroids of blobs belonging to the object in all frames. The same color of a track depicts consistent labeling and thus correct tracks. Notice the extremely small object sizes and the low contrast relative to the background. Background subtraction based methods fail in such scenarios where the lack of intensity difference between object and background result in a large number of false negatives (Fig. 19).

5 Conclusion

The chapter has presented a detailed analysis of the various steps in the aerial video tracking pipeline. In addition to providing an overview of the related work in the vision literature, it lists the major challenges associated with tracking in aerial videos, as opposed to static camera sequences, and elaborates as to why the majority of algorithms proposed for static camera scenarios are not directly applicable to the

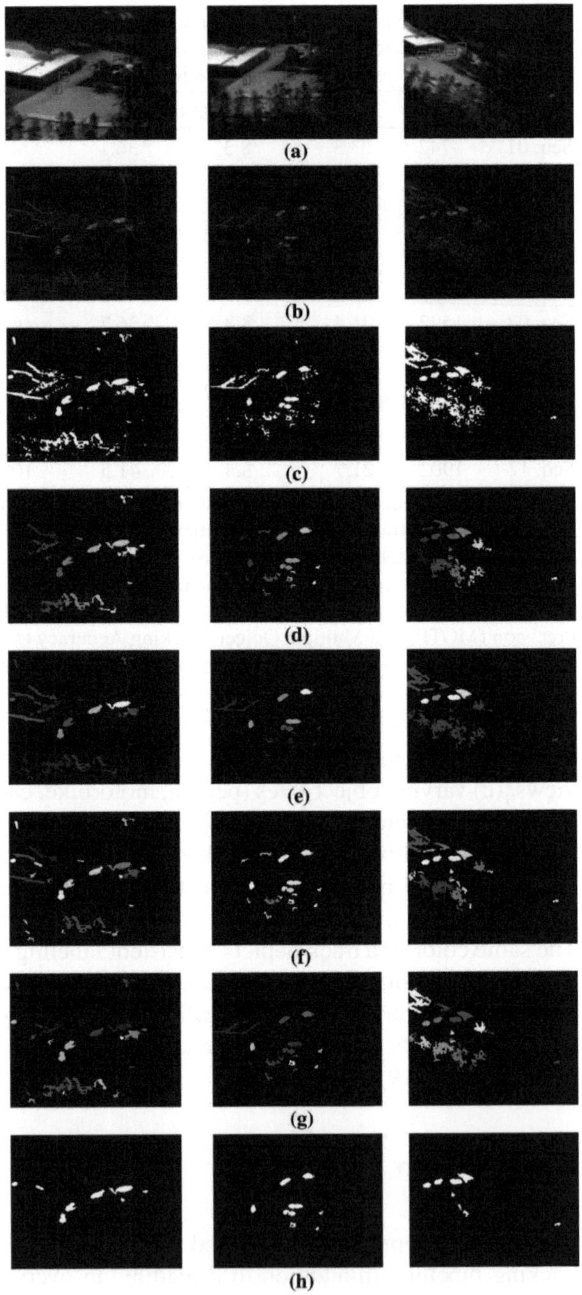

Fig. 19 Intermediate results for three frames from Distance View Sequence. *Bounding rectangles* in the original frames show the positions of groundtruth. **a** Original frames, **b** accumulative frame difference, **c** AFD > *T*, **d** connected components (30, 17 and 23), **e** mean gray area, **f** compactness, **g** eccentricity, **h** output blobs

aerial video domain. We have presented both the theoretical and practical aspects of a tracking system, that has been validated using a variety of infrared sequences.

References

1. Ali, S., Shah, M.: Cocoa—tracking in aerial imagery. In: SPIE Airborne Intelligence, Surveillance, Reconnaissance (ISR) Systems and Applications (2006)
2. Andriluka, M., Roth, S., Schiele, B.: People-tracking-by-detection and people-detection-by-tracking. In: CVPR (2008)
3. Arambel, P., Antone, M., Landau, R.H.: A multiple-hypothesis tracking of multiple ground targets from aerial video with dynamic sensor control. In: Proceedings of SPIE, Signal Processing, Sensor Fusion, and Target Recognition XIII, vol. 5429, pp. 23–32 (2004)
4. Bay, H., Tuytelaars, T., Gool, L.V.: Surf: speeded up robust features. In: ECCV (2006)
5. Berclaz, J., Fleuret, F., Fua, P.: Robust people tracking with global trajectory optimization. In: CVPR (2006)
6. Bernardin, K., Elbs, A., Stiefelhagen, R.: Multiple object tracking performance metrics and evaluation in a smart room environment (2006)
7. Blackman, S., Popoli, R.: Design and Analysis of Modern Tracking Systems. Artech House, Boston (1999)
8. Bouguet, J.: Pyramidal implementation of the Lucas–Kanade feature tracker: description of the algorithm. TR, Intel Microprocessor Research Labs (2000)
9. Brown, L.G.: A survey of image registration techniques. ACM Comput. Surv. **24**(4), 325–376 (1992)
10. Dalal, N., Triggs, B.: Histograms of oriented gradients for human detection. In: CVPR (2005)
11. Fischler, M., Bolles, R.: Random sample consensus: a paradigm for model fitting with applications to image analysis and automated cartography. Commun. ACM **24**(6), 381–395 (1981)
12. Gandhi, T., Devadiga, S., Kasturi, R., Camps, O.: Detection of obstacles on runway using ego-motion compensation and tracking of significant features. In: Proceedings of the 3rd IEEE Workshop on Applications of Computer Vision, p. 168
13. Heitz, G., Koller, D.: Learning spatial context: using stuff to find things. In: ECCV (2008)
14. Isard, M., Blake, A.: Condensation: conditional density propagation for visual tracking. In: IJCV (1998)
15. Jepson, A., Fleet, D., El-Maraghi, T.: Robust online appearance models for visual tracking. In: IEEE TPAMI (2003)
16. Kumar, R., Sawhney, H., Samarasekera, S., Hsu, S., Tao, H., Guo, Y., Hanna, K., Pope, A., Wildes, R., Hirvonen, D., Hansen, M., Burt, P.: Aerial video surveillance and exploitation. IEEE Proc. **89**, 1518–1539 (2001)
17. Leibe, B., Schindler, K., Gool, L.V.: Coupled detection and trajectory estimation for multi-object tracking. In: ICCV (2007)
18. Lin, R., Cao, X., Xu, Y., Wu, C., Qiao, H.: Airborne moving vehicle detection for video surveillance of urban traffic. In: IEEE Intelligent Vehicles Symposium, pp. 203–208 (2009)
19. Lowe, D.G.: Distinctive image features from scale-invariant keypoints. Int. J. Comput. Vis. **60**, 91–110 (2004)
20. Mann, S., Picard, R.W.: Video orbits of the projective group: a simple approach to featureless estimation of parameters. IEEE Trans. Image Process. **6**, 1281–1295 (1997)
21. Matas, J., Chum, O., Martin, U., Pajdla, T.: Robust wide baseline stereo from maximally stable extremal regions. In: BMVC (2002)
22. Muja, M., Lowe, D.G.: Fast approximate nearest neighbors with automatic algorithm configuration. In: VISAPP (2009)

23. Olson, C.F., Huttenlocher, D.P.: Automatic target recognition by matching oriented edge pixels. IEEE Trans. Image Process. **6**, 103–113 (1997)
24. Perera, A., Srinivas, C., Hoogs, A., Brooksby, G., Hu, W.: Multi-object tracking through simultaneous long occlusions and split–merge conditions. In: CVPR (2006)
25. Piccardi, M.: Background subtraction techniques: a review. In: IEEE International Conference on Systems, Man and Cybernetics, vol. 4, pp. 3099–3104 (2004)
26. Shah, M., Kumar, R.: Video Registration. Kluwer Academic Publishers, Dordrecht (2003)
27. Shi, J., Tomasi, C.: Good features to track. In: CVPR, pp. 593–600 (1994)
28. Spencer, L., Shah, M.: Temporal synchronization from camera motion. In: ACCV (2004)
29. Xiao, J., Cheng, H., Han, F., Sawhney, H.: Geo-spatial aerial video processing for scene understanding. In: CVPR (2008)
30. Xiao, J., Yang, C., Han, F., Cheng, H.: Vehicle and person tracking in aerial videos. In: Multimodal Technologies for Perception of Humans: International Evaluation Workshops CLEAR 2007 and RT 2007, pp. 203–214 (2008)
31. Yalcin, H., Collins, R., Black, M., Hebert, M.: A flow-based approach to vehicle detection and background mosaicking in airborne video. In: CVPR, p. 1202 (2005)
32. Yalcin, H., Collins, R., Hebert, M.: Background estimation under rapid gain change in thermal imagery. In: OTCBVS (2005)
33. Yilmaz, A.: Target tracking in airborne forward looking infrared imagery. Image Vis. Comput. **21**(7), 623–635 (2003)
34. Yilmaz, A., Javed, O., Shah, M.: Object tracking: a survey. ACM Comput. Surv. **38**(4), 1–45 (2006)
35. Yilmaz, A., Shafique, K., Lobo, N., Li, X., Olson, T., Shah, M.A.: Target-tracking in flir imagery using mean-shift and global motion compensation. In: Workshop on Computer Vision Beyond the Visible Spectrum, pp. 54–58 (2001)
36. Yin, Z., Collins, R.: Moving object localization in thermal imagery by forward–backward mhi. In: OTCBVS (2006)
37. Yuan, C., Medioni, G., Kang, J., Cohen, I.: Detecting motion regions in presence of strong parallax from a moving camera by multi-view geometric constraints. IEEE TPAMI **29**, 1627–1641 (2007)
38. Zhang, H., Yuan, F.: Vehicle tracking based on image alignment in aerial videos. In: Energy Minimization Methods in Computer Vision and Pattern Recognition, vol. 4679, pp. 295–302 (2007)

MIX
Papier aus verantwortungsvollen Quellen
Paper from responsible sources
FSC® C105338

If you have any concerns about our products,
you can contact us on
ProductSafety@springernature.com

In case Publisher is established outside the EU,
the EU authorized representative is:
Springer Nature Customer Service Center GmbH
Europaplatz 3, 69115 Heidelberg, Germany

Printed by Libri Plureos GmbH
in Hamburg, Germany